Time, Space and Innovation: an Archaeological Case Study on the Romanization of the North-Western Provinces (50 BC to AD 50)

Claudia Dürrwächter

BAR International Series 2011
2009

Published in 2016 by
BAR Publishing, Oxford

BAR International Series 2011

Time, Space and Innovation: an Archaeological Case Study on the Romanization of the North-Western Provinces (50 BC to AD 50)

ISBN 978 1 4073 0587 5

BAR Publishing is the trading name of British Archaeological Reports (Oxford) Ltd. British Archaeological Reports was first incorporated in 1974 to publish the BAR Series, International and British. In 1992 Hadrian Books Ltd became part of the BAR group. This volume was originally published by Archaeopress in conjunction with British Archaeological Reports (Oxford) Ltd / Hadrian Books Ltd, the Series principal publisher, in 2009. This present volume is published by BAR Publishing, 2016.

Printed in England

BAR
PUBLISHING

BAR titles are available from:

BAR Publishing
122 Banbury Rd, Oxford, OX2 7BP, UK
EMAIL info@barpublishing.com
PHONE +44 (0)1865 310431
FAX +44 (0)1865 316916
www.barpublishing.com

CONTENT

LIST OF FIGURES

CHAPTER 10

CHAPTER 11

CHAPTER 12

LIST OF TABLES

ACKNOWLEDGMENTS

I would like to take this opportunity to thank various people that have helped me and at times pushed me along this tremendous journey. I would like to thank my supervisor Nick Winder, for his guidance, encouragement and counsel during my Phd, and also for helping me find funding for the project. I could not ask for a more supportive and patient supervisor.

Thanks also go to Kevin Greene and Jeremy Paterson for their archaeological expertise and valuable suggestions and to everyone who helped me through my research: Tom Moore for his comments on early drafts of my work, all the researchers of TiGrESS, Ellen Hambleton and Mark Maltby for help with understanding the dataset. Special thanks also to Sheila Newton, for her assistance with editing the final versions of this manuscript.

My research was financed by TiGrESS (EVG1-2002-00081).

Finally I would like to thank my family – particularly my mum for giving me encouragement, emotional and financial support throughout my academic life, and providing a firm foundation for me to spread my wings and fly.

INTRODUCTION

"We have a somewhat less than perfect understanding of cultural dynamics" (McGlade & van der Leeuw 1997: 3)

This book is concerned with social stability and change. Despite continuing interest in both aspects by various disciplines of the social sciences they are still not fully understood. Unlike the natural sciences, where Darwin's principles of random variation and selection are commonly accepted as mechanisms of change, the social sciences still lack a paradigm of cultural evolution and the explanation of social change remains a crucial question.

During the 1960s and 70s almost all the social disciplines experienced attempts at applying so-called 'hard-sciences' methods to cultural change. Several approaches have tried to transfer Darwin's paradigm to social science problems (Wilson 1975, Dawkins 1976). They are all based on the assumption that human behaviour evolved in the same way as our bodies and that our understanding of biological change can be equally well applied to cultural evolution. None of these attempts has avoided severe criticism (Lewontin et al 1988, Ziman 2000, Read 2003). The reason for this failure is closely connected to the nature of the research objects and has given us valuable insight into the character of social dynamics. It has been concluded that the processes that constitute the world of nature are qualitatively different from the processes that determine human history. Traditionally, *hard science approaches* to human systems are based on the assumption that scientific laws about human behaviour can be treated as time-invariant. The dynamic systems that govern human behaviour have a fixed causal mechanism and differences in the socio-natural state observed through time and space are explained solely in terms of the operation of that system. *Soft science methods*, however, have to accept that human knowledge is socially constructed and, as that knowledge changes, so human behaviours actually change with it. Not only does the state of a soft system change through time, but the nature of socio-natural dynamics changes too. Social processes often create what have been described as "wicked problems" by Rittel and Webber (1973) in as much as no time-invariant objective solution exists for them; there is no general answer for questions like what is "beautiful" or "just" or causing criminal behaviour in young adults. The historian Collingwood (1946: 213-217) argued that historical processes have an outside as well as an inside while natural processes only have an outside. It is the 'inside' processes that add complexity to social phenomena and thus make them difficult to comprehend, explain or manage. This is due to the fact that social change is often non-linear, unpredictable and may result in emergent pattern and unintended consequences. The interaction between the micro and the macro level of society, or the link between individual behaviour and macro-scale structural processes, also remains far from being understood (McGlade & van der Leeuw 1997, Vickers 1995: 82-90, Schimanck 2000: chapter 1).

This study will focus on one particular mechanism of social change: *innovation*. It can be defined broadly as a cultural process that somehow changes our world, often resulting in an acceleration of history (Hann 1994). Innovation - by definition - describes qualitative rather than quantitative or gradual change. The phenomenon has been studied by a number of scientific disciplines with a range of purposes and this has resulted in many perceptions and definitions. The literature on innovation reveals a great variability in the understanding of its underlying mechanism, even within disciplines of the humanities, social- and policy-relevant sciences (Fagerberg 2005). Although innovation is generally considered to be an important aspect of social change no unambiguous definition exists to explain it, which is a significant disadvantage in regard to interdisciplinary research and communication.

In this book it will be argued that *conceptual modelling* provides a useful method to tackle ill-defined and complex social phenomena like innovative processes. Any modelling exercise forces the researcher to identify the crucial components in regard to a particular research question. A model should represent a sensible reduction of reality and thus helps to understand and operationalise its complexity. The application of modelling to the social sciences, originally used as quantitative representations of reality, has experienced the same difficulties as Darwinian theories of social change. Again it has proven difficult to capture human behaviour and its underlying intentions with concepts like rationality or linearity. The emergent nature of accumulated human actions disputes with the assumption that any system can be explained merely as the sum of its parts. Based on these difficulties the social sciences have tried to widen the concept of modelling to adapt it to the specific nature and requirements of cultural dynamics.

This research is associated with the European Union's V framework programme on Energy, Environment and Sustainable Development. With TiGrESS (Time Geographical Approaches to Emergence and Sustainable Societies) the EU has invested in a project developing Time- Geographical methods. Scientists from many different backgrounds are involved (economists, modellers, geographers) which gives TiGrESS the opportunity to combine human and technical resources from different European institutes and scientific disciplines which would not be possible for one department or university alone. Time-Geography is one of the few disciplines within the social sciences that stress the importance of both space and time in the understanding of social processes. Both factors are determined by human perception and at the same time define the opportunity space of individuals (Hägerstrand 1967, McGlade 1995). This background provides excellent conditions for interdisciplinary research.

A conceptual model of innovation that is based on the literature of various disciplines of the social sciences and equally respects their perspectives and understanding of that topic will be developed in this book. This model will respect the complex and often problematic relationship between the micro and the macro level. It will consider both the role of relatively short-term human behaviour as well as the long-term structuring of these behaviours in the reproduction of societies and thus connect to the sociological understanding of the human condition (Schimanck 2000, Giddens 1984, Archer 2005). It will also provide a better understanding of the underlying mechanism of innovation beyond the artificial borders of scientific disciplines.

The usefulness of the model will be tested by applying it to the archaeological record. The Romanization of the North-Western Provinces 50 BC to AD 50 is investigated using visual as well as statistical analysis on the faunal remains of cows, sheep and pigs. The choice of case study is a rather random one, because the model claims to be acceptable for all disciplines of the social sciences and could therefore be tested on any other innovation manifest in the archaeological record or any other discipline. Looking at human prehistory it is obvious that there have been cultural changes as well as innovations. One example is the transition from a lifestyle of hunting and gathering to agriculture and animal domestication, known as the Neolithic Revolution. Another crucial novelty is the development of urbanization and of state societies.

Due to the nature of the archaeological record and differential preservation processes altering the original picture, archaeologists work with incomplete documentation of human behaviour. They merely have artefacts and skeletal remains to reconstruct former cultures and ways of life. Therefore dynamic cultural processes must be reconstructed indirectly from static artefacts and skeletal remains. Consequently Binford (1983) called the archaeological evidence an "untranslated language". The archaeological record lacks any direct information about cognitive data and processes, which have to be reconstructed purely by indirect methods. Even though our knowledge about past processes is limited by these factors, a historical view has the advantage of allowing for an investigation of change over a long period of time. A successful application of the model to the archaeological record with all its limitations will support its usefulness for other, more contemporary, research questions which have better access to human behaviour.

This book is thus divided into two parts: the first one is concerned with *theory building* (developing a conceptual model of innovation) and the second part is engaged in *theory evaluating* (archaeological case study). Both processes are intentionally separated as much as possible, in order to minimize the circular argument trap. The data in which the conceptual model was developed (innovation literature) is completely independent from the data it is tested in (archaeological record). This approach tries to gain independence between the ideas and the evidence while finding a way to connect the conceptual model with empirical data.

This book will:

1. Find a definition of innovation that can be applied with equal facility in different branches of the social sciences namely: archaeology, social geography, economics and policy-research.

2. Explore the process of innovation in the archaeological record of Europe especially on the Romanization of the North-Western Provinces and its attendant social changes. The application of the conceptual model of innovation to the archaeological record will provide new insights into pre-historical processes as well as testing the definition's applicability for all four scientific domains mentioned above.

3. Extend techniques from Time Geography that have been developed in an EU funded project on time geography (TiGrESS – [EVG3-2001-00024]) to the study of innovation in the historical and archaeological record.

This is not an ordinary archaeological case study based on expertise in one area, but rather an attempt at truly interdisciplinary research. It tries to bridge the gap between quantitative and discursive methods as well as the boundaries of modern disciplines because it is felt that social change affects all aspects of human society and cannot be fully investigated from any one-sided perspective.

PART 1 – A CONCEPTUAL MODEL OF INNOVATION

Theory building

1. METHODS

This book is concerned with continuity and change in complex social situations. Innovations are an important mechanism of social change. Unlike the natural sciences, where Darwin's principles of random variation and selection are commonly accepted as mechanisms of change, the social sciences still lack a paradigm of cultural evolution and an explanation for social change. Attempts to transfer the Darwinian principles to the social sciences - e.g. sociobiology (Wilson 1975); meme theory (Dawkins 1976) or the dual inheritance approach (Boyd & Richerson 1985) - are still controversial. They have triggered substantial criticism and many promises have not been fulfilled. This can be attributed to the fact that even though there are basic parallels between cultural and natural evolution there are also many 'disanalogies' (Ziman 2000: 5). Efforts to transfer Darwin's principles to cultural change were not the only attempts to apply natural science methods to social science problems. Disciplines like management and policy science or operational research went down the same route. Archaeology also showed a growing interest in science and scientific methods, and computer simulations became very fashionable in the 60s and 70s.

All these attempts experienced problems which triggered debate and search for solutions during the 1970s. The same explanation was given in all disciplines: the realisation that there are two qualitatively different kinds of problems in the world and natural science methods are only suitable for one of them (Rosenhead & Mingers 2001, Schecter 1991). The variety of terms used to classify these problems - 'objectivist' versus 'subjectivist' (Rosenhead & Mingers 2001, 'wicked' versus 'tame' (Rittel & Webber 1973), 'soft' versus 'hard' (Checkland 2001) to name just a few – might camouflage the idea that they all describe the same basic problems.

Natural science or engineering problems are generally solved by bridging the discrepancy between the status quo and a desired state in the most cost-effective and elegant way. The problem is clear and well-defined right at the beginning of the research. A classic example for this type of problem is the American moon landing, where the objective was clearly defined by Kennedy himself as "before the decade is out ... landing a man on the Moon and returning him safely to Earth" (quoted after Checkland 2001: 62). It is also possible to assess whether a problem was solved or not. Natural science problems can therefore talk about "problems" and "solutions" which abolish problems (Checkland 1991).

The "real-world" problems of the social sciences are very different in as much as they are messy and ill-defined. Social systems are closely intertwined and the outcome of one system might become the input into another system. The pluralities of stakeholder interests within a group make it very difficult for a policy maker or social scientist to pursuit unitary aims. While a certain degree of human consensus is crucial for the existence of institutions, there will never be a complete overlap of perception and evaluation. This problem connects to another insufficiency that arises with the transfer of system engineering methods to social systems: the assumption that the concept of 'goal-seeking' can be used to explain human behaviour. The definition of goals has proven difficult in real-life situations and Vickers has argued that it is the maintaining of relationships rather than goal-seeking that determines human activities (see a more detailed discussion in chapter 3). Decisions about which goal is more important within the plurality of interests are very difficult and closely related to questions of power (Checkland 1991, 2001).

Social science problems make it very difficult to define where the problem centre lies. Therefore to *define the problem* and "knowing what distinguishes an observed condition from a desired condition" (Rittel & Webber 1973: 159) is the actual problem within the social sciences. The information gathered to identify and understand the nature of a wicked problem will also influence the understanding of its solution. For instance, if the lack of education is identified as the source of criminal behaviour in teenagers, the obvious way to tackle this problem is to improve education. If we think teenagers become criminals because they grow up in a difficult family situation we would aim at improving this. "Problem understanding and problem resolution are concomitant to each other" (Rittel & Webber 1973: 161). Once we understand why there is a difference between what is and what should be we automatically know what to do in order to overcome this discrepancy.

The problem definition depends on the researcher's knowledge and experiences because these factors influence which explanation appears most plausible to her or him (see chapter 3). Therefore two different people looking at the same problem might develop very different solutions. In contrast to the natural sciences there is no way of telling which of these solutions is better for a particular problem.

Vickers' concept of *appreciative systems* can help to make this point more explicit. He (1968: chapter 7) describes the human decision-making process as *appreciation* and claims that humans make three kind of judgments: reality, value and operational. *Reality judgements* are basically schemata of reality, our definition of *what* is. They are based on the present and the past and thus depend on an individual's memories and experiences. *Value judgments* define how we want reality to be. These judgments cannot be proven correct or incorrect. In the decision-making process humans compare what is (reality) with what they want to be (value) and decide which action to take in order to overcome discrepancies between the two. These actions are based on *operational judgements*. However the last step may be irrelevant as an appreciation may or may not call for action. The appreciative system changes with every new piece of information an individual processes and this process is generally gradual and unconscious.

Every human decision is based on appreciation, including professional ones. Decisions about the nature of a social problem are therefore influenced by the personal value and reality judgments of the researcher. Clarke (1972) makes a very similar point when he differentiates between two different kinds of model: operational and controlling (see figure 1). While operational models stand for the methods used by a researcher, controlling models represent his beliefs and thus influence his behaviour including scientific choices. Clarke stresses that humans always use conceptual models to make sense of their environment and scientists cannot free themselves from this mechanism in their professional lives. The history of science is full of examples that show how beliefs have influenced the interpretation of data. The reading of the Taung child, a juvenile early hominid first published in 1925 by Dart, gives a good example. Surrounding broken animal bones and deposits were originally interpreted as australopithecine tools and weapons, described as the 'osteodontokeratic culture' (bone, tooth and horn), a picture which fitted into the 'man as hunter' hypothesis valid at the time. This interpretation was later rejected as unrealistic, because a hominid of such a small size and stature could not have killed its prey. There is always the danger that we "observe what we believe and then believe in that which we have observed" (Clarke 1972: 6).

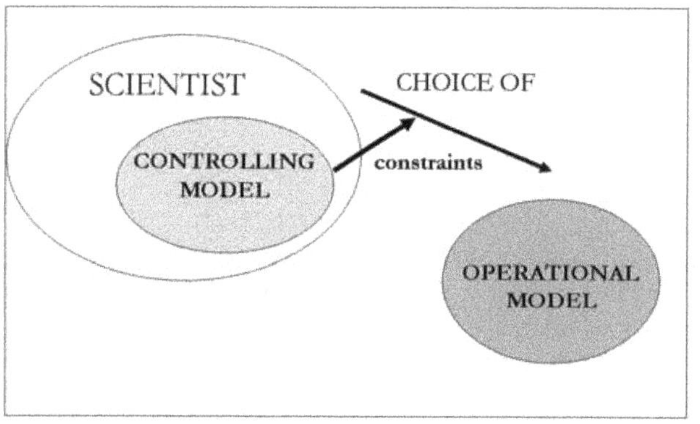

Figure 1: *Clarke's (1972) understanding of scientific decision making; embodied controlling models constrain the conscious selection of conceptual models used to explain reality and thus interfere with our perception of scientific problems*

Another aspect of wicked problems is that any action taken actually changes the problem situation. Therefore an attempt to solve this kind of problem is a 'one shot operation' and leaves irreversible traces. If it leads to unforeseen and unwanted consequences, attempts to reverse the action are faced with another wicked problem (Rittel & Webber 1973, Checkland 2001).

Checkland has argued that hard system methods might be suitable for some – well defined - social science problems but in most cases they are too complex to be captured as the efficient achievement of goals engineered as problems and their solution. "[...] the relation between "hard" and "soft" systems thinking is not like that between apples and pears: it is like that between apples and fruit" (1991: 71).

1.1 MODELLING

This brief description of the nature of social science problems should have demonstrated why they cannot be expected to easily fit into Darwinian evolution or reduce naturally into any quantified form. Therefore methods developed for the natural sciences cannot be expected to help with their solution. The concept of problems and their solution, computer simulations or hard systems models are only applicable for well-defined and bounded problems and the social sciences need their own approaches to address the complex and messy problems they are dealing with. Rosenhead and Mingers (2001) differentiate between *problem solving* and *problem structuring* methods. The latter uses models with little or no quantification, so-called conceptual models (see figure 2). The complexity and pluralistic nature of reality studied by the social sciences requires discursive and dialogical methods (McGlade 1995).

Conceptual models provide a valuable tool to simplify and operationalise complex social phenomena. They are helpful in structuring our understanding of a problem. Checkland has argued that models should be regarded as models relevant to arguing about the world but not models of the world (Checkland & Caspar 1986, Checkland 1991, 2001). The process of building any model - a computer simulation or a conceptual model - starts with defining which components are crucial in regard to the research question. This process is essential as social science problems are too complex to be fully described. This first step forces the researcher to clear his mind and to identify the crucial components in regard to the research problem.

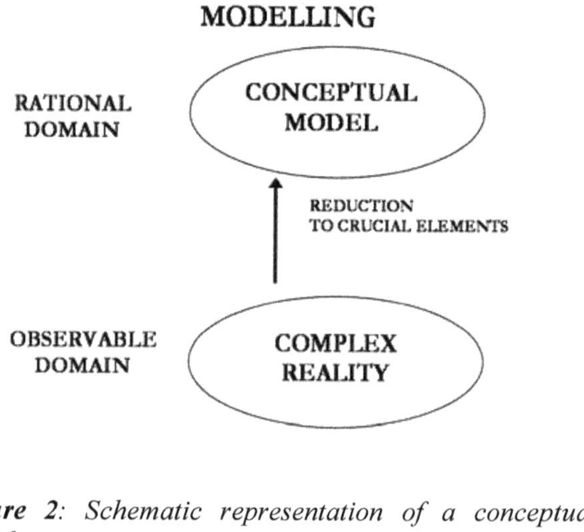

Figure 2: *Schematic representation of a conceptual model*

Models do not claim the ability to copy real life situations; reality is far too complex to be copied in any model. A model does not represent reality but rather our *understanding of reality*. As Bonabeau et al. put it "a model is a simplified picture of reality: a usually small number of observable quantities, thought to be relevant, are identified and used as variables; a model is a way of connecting these variables" (1999: 17). They simplify reality by the "selective elimination of detail incidental to the purpose of the model" (Clarke 1972: 2, Kohler 2000, Checkland 2001). It is, in fact, the careful selection of few factors according to a research question that account for a model's utility and the information that can be gained from it (see figure 2). Clarke defined models as "pieces of machinery that relate observations to theoretical ideas" (1972: 1).

Nevertheless any useful model must represent the behaviour of real people in a real environment in a meaningful way. And a conceptual model must be compared with real-world action. A model's usefulness will be judged by others through their acceptance of this selection and their agreement that all relevant aspects of the problem have been captured (Checkland 1991, Kohler 2000).

Even a very brief occupation with innovation shows that it is complex, involves many factors and is nonlinear (Kline & Rosenberg 1986). Non-linearity means that innovative processes do not follow "smoothly down a one-way street" (ibid: 285) with mechanical causality but are characterised by feedback loops, evolutionary potential and uncertainty. Complexity implies that innovation is irreducible and more than the sum of its parts (McGlade & van der Leeuw 1997). This book will therefore address the study of innovation by developing a conceptual model. This model will be based on the understanding of innovation (reality judgements) of various disciplines of the social sciences. It will represent the mechanism behind innovative processes or in accordance to soft system theory it will make "plain its *Weltanschauung*, the point of view from which the (human activity) system is described – since one man's "terrorism" is another's "freedom fighting" (Checkland 1991: 68).

The next two chapters will gather information that will provide the basis for such a conceptual model.

1.2 TIME-GEOGRAPHY

"In almost the same manner as the world-pictures of technologists and biologists have been empty of people, the scenarios entertained by humanists and social scientists, including historian, have neglected the role of natural conditions and artefacts" (Hägerstrand 1987: 47)

Human behaviour plays a crucial part in many aspects of social change. However the framework under which human behaviour is studied and explained varies throughout the social sciences. Time-Geography (TG), developed by Torsten Hägerstrand in the 1960s, has focused on the importance of *time* and *space* in the constitution of the human condition and thus differs from many other social science disciplines. It sees human biographies as an uninterrupted succession of time-uses and time and space as intimately related (Hägerstrand 1967, 1987, 1988). Both factors have been neglected by most other approaches to social change or studied in separation from each other. The rational-choice approach for instance acknowledges time as a limiting factor on human behaviour but ignores the physical landscape (Becker 1981, Esser 1990). Constraints like the fact that humans cannot be at two places at a time or that any task consumes time appear to be obvious but Time-Geography argues that they are nonetheless crucial to the explanation of human behaviour because they represent constraints. Every purposive human action has to consider its limitations.

TiGrESS set out to apply Hägerstrand's methods to complex adaptive systems, which required some alterations. Traditionally the focus in Time-Geography was individual-based and models constructed from the micro-level. However while many problems of the social sciences demand local solutions, they also require policy regulations on a higher level of the community. Thus a solely micro-level approach is problematic and TiGrESS investigated the usefulness of a *multi-scalar* approach to human space time behaviour, which allows the observation of 2 or 3 organizational levels simultaneously (Pumain & Favaro 2006, Winder 2006). Sanders (1999) has argued that diversity at the micro level can be represented with stochastic components at the meso-level (see also Haken 1978). This so-called synergetic approach is based on the assumption that individual behaviour only creates a simple and more or less random difference which is determined through meso-level processes like economic or social profiles of an area. Rather than aggregate individual behaviour and decisions and emphasise the individuality of the actors this approach uses statistics to represent individual behaviour. Hägerstrand's Time-Geography also focuses on human behaviour but does not provide us with methods to observe human beliefs. TiGrESS aimed to develop a conceptual model about the impact of the human perception on the landscape. The environment is never neutral but rather shaped by the relationship that people living in a certain area have with it. The concept of space is thus always related to human perception (McGlade 1995). This aspect of TiGrESS links TG to discursive approaches.

2. INNOVATION: AN ILL-DEFINED MECHANISM OF CHANGE

The term "innovation" has become so fashionable it is hard to avoid: Economists talk about innovation as an important factor in economic growth and competition (Freeman 1982, Sundbo 1998, Edquist 2000, Dosi 2000). Politicians express a similar view, emphasizing the need for innovation to guarantee a better future (EU 1995). Historians point out how major innovations in the past changed the course of history and continue to influence our lives. Cultural Geographers study the diffusion of innovations as a spatial process (Hägerstrand 1967) while the advertising industry considers even the smallest variation in a product an innovation that will improve our lives remarkably if we only buy it.

One aspect that is shared between all these perceptions is that innovation is a complex process involving numerous - often unidentified - factors. The observable variation within innovation theory should therefore not be surprising. But are all these people talking about the same thing? Or does the term innovation – simultaneously used to describe a new mobile phone technology and novel forms of social organization – cover very different kinds of phenomena? Is innovation an ill-defined term? These questions might appear redundant at first because every discipline, taken in isolation, seems very confident about what innovation is and how it should be managed. But a closer look at the literature discloses the existence of many different classifications of the phenomenon, even within one research area. This variation widens between different disciplines. The importance that is ascribed to innovative processes today, combined with an apparent wish to stimulate and control this phenomenon, will make interdisciplinary discussion inevitable and a non-ambiguous definition of what innovations actually are is crucial for these processes. Such a definition might additionally provide understanding of the mechanisms underlying innovative processes, which is indispensable for a meaningful management of complex social phenomena. The acclaimed social significance of innovation also adds ethical dimensions. This chapter is to give an overview of the present discussion on innovation. Because of the abundant literature on the subject (Fagerberg 2005) it is impossible to exhaust every line of argument within it and this research focuses only on the social sciences, covering disciplines with mainly contemporary issues that express a strong wish for ways of controlling and predicting innovative processes (politics, economics) as well as research with a retrospective view (archaeology, cultural geography). The following literature review will summarize the research interests of each approach in order to clarify their line of argument and intention. Within the available literature, economic publications are by far the most abundant.

2.1 INNOVATION, INVENTION AND DIFFUSION

Innovations are often considered as synonymous with anything "new" but a closer look at the literature reveals that the term is not as clear-cut as it might seem. Does an innovation have to be something radically new that has never been seen in the world before or does it only have to be new to the unit as adoption as Rogers and Shoemaker (1971: 19) define it? What about incremental change? Or an innovation that was developed in one context and later implemented in another? Schumpeter introduced the term "imitator" for the latter case but it could be argued that this is an innovation as well because it introduces the innovation for the first time within a new context (Fagerberg 2005).

Innovation is often separated from closely related terms like creativity, invention and diffusion. Inventions are usually seen as the initial idea for a new product or process while innovation is the first attempt to implement an invention (Freeman & Soete 1997: 6, Stoneman 1995, Sundbo 1998: 19, Dosi 2000, Fagerberg 2005). An invention can be an idea or model of a new product, system or device. It only becomes an innovation in the economic sense if it gets transformed into a product or device that is actually involved in commercial transactions. "Innovation is concerned with the commercial and practical application of ideas or inventions. Invention, then, is the conception of the idea, whereas innovation is the subsequent translation of the invention into the economy" (US Department of Commerce 1967, quoted in Trott 2002: 12). A distinction between an initial idea and its actual realisation can also be found in the archaeological literature. Van der Leeuw and Torrence, for example, define inventions as: "the original conception of a new idea, behaviour or thing" while the word innovation describes "a continuous process consisting of a number of different components" (1989: 3). Both can be closely interlinked or separated by a time lag (Trott 2002: chapter 1, Stoneman 1995, Fagerberg 2005, Smith 2005). Economists would argue that this time lag is caused by the conditions for commercialisation which are not yet ready, while other social scientists would focus on cultural aspects of acceptance and resistance. There is a clear time line in this perspective, with an invention preceding an innovation. However invention and innovation are continuous processes and often cannot be separated (Fagerberg 2005).

The early economist Schumpeter (1939: 84-86) separates innovation from invention in a different way (see figure 3). He acknowledges that the latter creates new possibilities and is therefore an important cause of change. However he rejects the role of inventions in economic processes. He argues that the study of inventions would distract from the important innovation. For Schumpeter inventions and the corresponding innovations are entirely different things which underlie different processes. An invention and the means by which it is turned into an innovation differ even if they are carried out by the same person. "*Innovation is possible without anything we should identify as invention* and invention does not necessarily induce innovation, but produce of itself [...] no economically relevant effect at all" (1939: 84). By arguing that innovative processes are the same whether they are initiated by new or existing knowledge Schumpeter separates the two terms as well as rejecting the importance of knowledge creation stressed in the majority of the literature. He anticipated some of the problems that are experienced with economically exploiting research results today. "[...] it is the application of innovations (diffusion) rather than the generation of innovations (invention or research and development) that leads to the realization of benefits from technological advance (Stoneman 1991: 162) or in Schumpeter's own words: "As long as they are not carried into practise, inventions are economically irrelevant" (1949: 88). Schumpeter's perception varies from the rest of the literature which makes the same differentiation but does not exclude inventions from their research questions. Inventions are seen as the initial idea, while innovation is its actual diffusion or its transfer into business practices. While not every invention spreads to become an innovation, all innovations are seen as based on inventions (see fig 3).

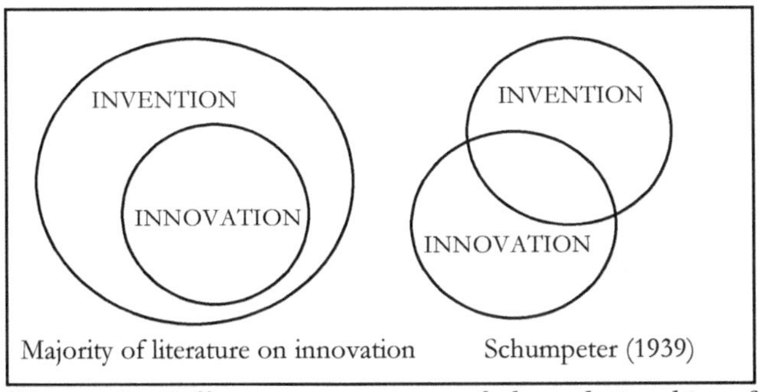

Figure 3: *Different perceptions of the relationship of invention and innovation*

The word "innovation" has been used within all disciplines to describe the whole process as well as its result (Freeman & Soete 1997: 6, EU 1995, Johnston et al. 2000: 175-177) and it has been argued that this differentiation is often blurred in practice (Dosi 2000). Both aspects provide a potential source of confusion.

The definition of innovation as a putting into practice of an invention also blurs the lines between innovation and diffusion. There seems to be a general concurrence about innovation being a complex process rather than a single event and the need for an initial invention or idea to spread within a group in order to be of any cultural or economic significance. That makes it difficult to pin down and adds a dynamic component. As Hall (2005) notes, the diffusion process does not only introduce an innovation throughout a population it is also an *intrinsic* part of the process which alters and often enhances an original innovation. Innovation can not be seen as a 'thing' but must be considered to be a process, which changes over time. An innovation at the end of the diffusion process might be significantly different from the innovation before and during the diffusion - changes that may, and often do, totally transform their economic significance. The subsequent improvements in an invention after its first introduction may be vastly more important, economically, than the initial availability of the invention in its original form" (Kline & Rosenberg 1986: 283).

The aspect of innovation diffusion carries essential questions in innovation research. Which factors decide if an invention can spread and is adopted by a significant number of people or organizations? The degree to which we can understand the mechanisms involved will determine our ability to influence and guide innovative processes. As demonstrated earlier, processes of social change are complex and interlinked with various other processes. The identification of the factors involved and their causal relationship often depend on the reality judgement (Vickers 1995: chapter 4), the research question (Rosenhead & Mingers 2001, Clarke 1972) and the boundary critique (Midgley et al 1998). The aspect of diffusion will be discussed in section 2.3.

2.2 INNOVATION AS A GUARANTEE FOR A BETTER FUTURE: THE ECONOMIC/ POLITICAL PERSPECTIVE

"... not to innovate is to die" Christopher Freeman (1982)

The macro economic and political literature on innovation often defines innovation as crucial for the development of a nation and its economy. While social and technological changes are seen as closely interlinked, the focus remains almost exclusively on the technological side (Dosi 2000) and innovations are studied in regard to their impact on economic growth and market competition. These approaches therefore favour innovations that have a direct impact on economic processes and produce changes observable on the macro level (BCG 2003, Dosi 2000, Sundbo 1998: 1; Stoneman 1995, EU 1995: 9). Because economic growth is often linked to a country's general well-being and important political issues like employment, the evaluation of innovative processes is very similar between politicians and economists. Both concentrate on the economic consequences of innovation and express a strong desire to manage and stimulate future innovative processes. A survey by the Boston Consulting group has shown that only 9.4 % of the companies questioned worldwide did *not* consider innovation as a top priority (2003). Corresponding to this view a significant amount of money is spent on identifying the crucial aspects involved in innovative processes in order to stimulate and manage them (Feldman 1994: 4 & 17). It is hoped that once these factors are identified, they will enable us to manage and control innovative processes. In 2000 the Lisbon agenda was launched with the aim of making the EU the world's most competitive economy by 2010, innovation was seen as an important factor for this process. But already halfway through the agenda it became apparent that the targets could not be met (EU 2005).

One of the aspects that have been identified as playing a key role in innovative processes is knowledge creation. This is hardly surprising considering that innovations are generally seen as synonymous to something "new" which either originated from new ideas or a novel combination of existing knowledge. New knowledge therefore provides the potential foundation for - successful - product innovation. According to this point of view investment in knowledge creation provides a means of supporting and encouraging innovative processes (Feldman 1994: 1, Stoneman 1995, EU 1995: 9). Consequently government and business funding are the two main sources of R&D (Patel & Pavitt 1995).

The aspect of knowledge is closely linked to the diffusion of innovation, where different types of knowledge have been identified. While *explicit* knowledge is generally considered as a "free" good that can easily be passed on or imitated (Polanyi 1998, Cohen & Levin 1989, Howells & Roberts 2000, Patel & Pavitt 1995) some have argued that the diffusion of *tacit* knowledge is influenced by geographical distance. Because the latter is not formally expressed but embodied, its transmission depends on human interaction and this type of knowledge does not 'travel' easily. Its nature defies easy articulation and makes it difficult to exchange over long distances (Polanyi 1998, Asheim & Gerther 2005). Geographers consider tacit knowledge as the key determinant of the diffusion of innovation. However research into knowledge spillover has shown that industries that have a high dependency on R&D and skilled labour have a tendency to cluster geographically, because they can better exploit the explicit knowledge which is created within spatial proximity. Phenomena like Silicon Valley support this theory (Jaffe 1986, Acs et al. 1992, Feldman 1994: chapter 2, Audretsch & Feldman 1996).

Economists generally consider only a narrow range of all available knowledge. Their focus on (macro-) economic processes excludes knowledge that is not related to creating and exploiting products or optimizing means of organization or regulation. Two major aspects of knowledge can be identified within the economic literature in regard to innovation: the creation of new knowledge and its economically successful diffusion (Dosi 2000, Audretsch & Feldman 1996, Howells & Roberts 2000).

In summary, the macro economic and political perspective assumes a causal relationship between knowledge creation, innovative activity and economic growth, which adds up to the following equation: knowledge generation triggers innovation which then leads to economic growth. While the nature and direction of this relationship have been subject to controversy and are still not fully understood, the belief in its fundamental connection remains unshaken (Dosi 2000). Reality of course has proven to be more complex. By focusing on technology development and product innovation, this macro perspective can only account for a small sector of the complexity in innovative processes. Even though the EU acknowledges the role of cultural aspects in the creation and acceptance of innovations they do not translate this into their policy statements (1995). Research funded by the EU is often aimed at either the creation of exploitable products or organizational improvements that help to increase a country's or company's innovative potential. While cultural factors, like a society's evaluation of risk takers or new technologies, can influence innovative processes significantly (EU 1995, Dosi 2000, von Pierer 2002), they are usually ignored and little money is invested in the research focused on this cultural side of innovations. Unsurprisingly this has created difficulties and the conversion of science into commercially exploitable products has been problematic (HoL 1997, Seaton & Cordey-Hayes 1993, Tomes 2003, BCG 2003). However this dissatisfaction did not result in a rejection of the assumed relationship between knowledge, innovation and economic growth, which is still considered as a "fact" in the economic literature. Rather it led to a revision of the nature of this relationship and discussion about its causal direction as well as the crucial factors involved (Dosi 2000). Observable failure is often credited to an inefficient transfer of research results (HoL 1997, Seaton & Cordey-Hayes 1993) or the tacit nature of knowledge. The paper on "Barriers to Innovation-Exploitation" by the House of Lords Select Committee on Science and Technology describes this problem as "the familiar refrain: strong in invention, weak in innovation" (HoL 1997: 3.6).

While a survey of global companies in 30 countries showed that 56.8 % were not satisfied with the financial return of their investment in innovation, about the same percentage stated that they nevertheless plan to increase their spending on innovation for 2004 (BCG 2003). Where does this imperturbable trust in a close relationship between innovation and economic growth come from? Why invest in a phenomenon when the desired effects cannot even be measured satisfactorily (Jaffe 1986, Basberg 1987, Acs et al. 1992, BCG 2003)?

2.3 INNOVATION AND DIFFUSION: GEOGRAPHY AND SOCIOLOGY

Cultural geography studies innovation as visible features in the landscape, which includes the investigation of its spatial origin as well as its spread from the area where it first occurred (Hägerstrand 1967: chapter 1). Human Geographers focus on the geographical diffusion of innovation and consider it as a spatial process. Hägerstrand identified two main modes of innovation diffusion: *contagion* and *hierarchical* diffusion. The first type proceeds through direct contact between individuals and produces spatial homogeneity (Pumain & Favaro 2006). The second diffusion mechanism follows a certain pattern; within a hierarchy it always starts at the high-level centres and diffuses down to the lower-levels as well as from the centre to the regional hinterland. This gives large cities an initial advantage in the adaptation process (Berry & Kasarda 1977: 279).

Beyond the macro level of economic growth and competition, economic research also considers individual factors like consumer behaviour, workplace environment, perception and communication as crucial to the innovation process (Moldovan & Goldenberg 2004). An innovation can only have a significant impact on the economy or people's lives if it spreads. Independently of the scale on which innovation is observed, its diffusion is always based on the interaction between individuals (Rogers & Shoemaker 1971: 23-24) and includes two sorts of actors: an advocate of change and a potential adopter (Fliegel & Kivlin 1966). The decision upon the acceptance of a new product or idea is therefore always placed within the domain of individual perception, which is clearly influenced by cultural structures. When confronted with something new every individual has to decide whether to accept or reject it. Such decisions are often based on an evaluation of the assumed advantages and disadvantages of the particular invention (Janis & Mann 1977: chapter 7). It has been shown that word-of-mouth communication can significantly influence this process in both directions (Hägerstrand 1967: 158-164, Moldovan & Goldenberg 2004). This perspective highlights the aspect of "trustworthy" rather than "efficient" in knowledge transfer relevant to decision processes. It supports the discussion on the diffusion of tacit knowledge.

It has been suggested that individuals differ in their willingness to accept innovations as well as in their tolerance towards uncertainty. It is generally accepted that innovations are not accepted instantaneously by an entire group and their diffusion process is never instantaneous. This can be observed in time-lags between the first occurrence and a broader diffusion (Hägerstrand 1967: chapter 6, Rogers & Shoemaker 1971: chapter 5, Allen & McGlade 1986, 1987, Dosi 2000). While some individuals exhibit a stronger willingness for taking emotional, financial or intellectual risks others choose to continue in their approved ways. By doing so, they guarantee continuity within society. Because absolute social change is impossible, some areas of society always have to reject change if it is not to be dissolved in chaos (Ellen 1994). On the level of society a general distribution pattern has been identified (McGlade & McGlade 1989, Allen 1989, Rogers & Shoemakers 1971: chapter 5). After acceptance by a small group of initial adopters the innovation is adopted with some delay by the large majority while a small group of individuals will principally reject it, producing a bell-shaped pattern of adoption. Allen and McGlade (1986) have argued that every society needs two different types of people: "Cartesians", individuals exploiting existing knowledge and keeping to traditional ways as well as "Stochasts", people that take risks by seeking out the unknown. While it is the latter group that drives change and the creation of new knowledge, any sustainable society depends on a dynamic balance between these two behavioural groups, as change can never grasp a whole society at once (Allen & McGlade 1987, Ellen 1994).

The question of what causes such differences in individual behaviour leads back to the individual level, with economic, social, geographical or demographic attributes. While some people have claimed that individuals can be generally classified as potential adopters or rejecters, Allen and McGlade (1986, 1987) argue that every individual can act in one way or another in regard to different circumstances.

Archaeologists have also emphasized the role of population density for the diffusion process. It has been argued that an increase in population density provides more innovative potential and thus better odds for successful innovation. More people are associated with elevated possibilities of knowledge transmission, creating a higher likelihood of innovation occuring (Hägerstrand 1988, Fitzhugh 2001). This argument has been used to explain the emergence of modern human culture with its astonishing innovations as linked to a significant population increase about 60 thousand years ago. From this perspective it is human potential rather than available knowledge that triggers innovative processes (Shennan 2001).

Innovation diffusion has also been of great concern in the macro economic literature, but again the focus here is more of an aggregated perspective than research on individual factors and interaction. The interest is based on the assumption that only the actual diffusion of a new technology or product can lead to economic success. While this approach also allows for differences between individual agents, these are often aggregates above the individual level, e.g. companies differing in size or research activities. The decision process of such agents is often described as a rational choice process, wherein an innovation is accepted if the associated costs are considered smaller than the benefits. Observable differences in the adaptation process are explained with imperfect information or differences in structural characteristics (Dosi 2000). There is evidence however that some innovations that had already been successfully implemented and proved to be beneficial were abandoned, e.g. firearms in Japan in the 1600s, as well as the diffusion of some innovations despite the existence of superior alternatives. The QWERTY typewriter, for instance, became the most common layout for keyboards even though people trained on another layout could type more quickly (David 1985). This is due to the fact that the diffusion process is path-dependent, which can lead to the adoption of the inferior version because conditions created by earlier decisions make that version more attractive (Hall 2005). This shows that cost benefit calculations are not the only determinants of the diffusion process. Traditional economic models of innovation diffusion account for this by assuming institutionalized behaviour rather than rational choice, but their focus on a descriptive approach provides no deeper explanation for the observable patterns (Dosi 2000).

2.4 INNOVATION AS A SOURCE OF CULTURAL CHANGE: A HISTORICAL PERSPECTIVE

In archaeology innovation is seen as one of the main causes of social change and the archaeological record clearly indicates that innovations changed the world throughout prehistory. Major changes like the introduction of agriculture, urbanisation or the invention of bronze and copper metallurgy have left traces in the archaeological record. One example of such extraordinary change is the transition from a lifestyle of hunting and gathering to agriculture and animal domestication, known as the Neolithic Revolution. These transformations occurred independently in different places of the world, one of them being the Fertile Crescent (Middle East) where we find the first visible traces of large-scale farming and animal husbandry around 10,000 years ago. These new ways of subsistence spread east into Europe until they finally reached Great Britain around 4000 B.C., resulting in people becoming permanently sedentary and higher population densities. Another crucial novelty that is visible as a revolutionary change in the archaeological record is the development of urbanization and of state societies. While traces of the first cities can be found in Mesopotamia as early as 9000 BC, with Jericho, and 7000 BC Catal Hüyük, there is no evidence for urbanization spreading into Europe until much later. European farmers lived in relatively small settlements and large cities are rare up to the early Iron Age.

Innovation is seen as "a continuous process consisting of a number of different components" (van der Leeuw & Torrence 1989: 3). Bargatzky defines innovation "as something which affects the overall performance of a sociocultural system at a certain time and which did not exercise this influence before - be it an idea, an object, or just a new inter-relationship between objects" (1989: 17).

While the macro economic and political perspective focuses on future processes and aims at economic profit, approaches in archaeology and cultural geography usually follow a retro perspective and try to find explanations for observable innovative processes in the past. Archaeologists study cultural changes in the archaeological record which, due to its nature and taphonomic processes, can only give an incomplete picture of past behaviours. Revolutionary changes that result in qualitative shifts in the type and visibility of the archaeological record are easier to observe than small-scale transformations (van der Leeuw 1989). As Flannery puts it: "It is vain to hope for the discovery of the first domestic corn cob, the first pottery vessel, the first hieroglyphic, or the first site where some other major breakthrough occurred. Such deviations from the pre-existing pattern almost certainly took place in such a minor accidental way that these traces are not observable" (1968: 85)

The retro perspective led to the identification of factors other than knowledge creation and economic growth. Even though archaeologists, like economists, study innovation in connection to technological change (Renfrew 1986, Ottoway 2001), because these processes are observable in the archaeological record it is apparent that their research interest differs substantially. According to the archaeological record cost benefit deliberations or economic profit have not been central for the diffusion of many new technologies in the past. Knowledge needed for the development of life-improving technologies was often available a long time before the invention actually became significant (Renfrew 1986). Therefore factors other than pure know-how must have played an important part.

Archaeologists see cultural factors as crucial for the course of innovative processes (Lemonnier 1993). Shared cultural beliefs of a population influence the ways in which technology is perceived as well as which usages appear appropriate. It is therefore not only the nature or "purely material" aspect of an invention that determines the way it is used or its degree of acceptance, but the way it is perceived. This perspective claims that observable changes within the archaeological record should not be studied without considering the cultural framework in which they took place (Guille-Escuret 1993, van der Leeuw 1993, Renfrew 1986, Ziman 2000).

A retro perspective has also been adopted by part of the economic literature. Studies of past innovative cycles have considered the national systems under which these changes took place, e.g. the industrial revolution in Britain, Germany and late 20th century Japan, and have shown that the structural environment can significantly influence a nation's innovative performance (Freeman & Soete 1997: chapter 12, Edquist 2000, Dosi 2000). The role of cultural factors had to be acknowledged and was transferred into a future perspective. These results have been used to predict which circumstances would be favourable for new innovations to take off and how these can be created. The management policies emerging from this perspective often focus on changes in the educational system (EU 1995). However the reliance on the advice of researchers working in the cultural domain, e.g. psychologists, sociologists or archaeologists, appears to be small. Furthermore differences in goals – explaining the past as opposed to regulating the future – results in logically irreconcilable conceptual models.

Unlike the macro-economic and political perspectives a retrospective view can demonstrate the effects of innovations on different temporal scales (Braudel 1980: 25-54, van de Leeuw 1989). Economists are interested in proximate pay-backs of innovative activities and concentrate on the so-called *l'histoire événementielle,* the history of short time events (Braudel 1980: 34). However the social consequences of innovations often reach out further, are hardly visible and of far greater significance to people's lives than the impact of economic growth. The introduction of the mobile phone, for instance, with its resulting changes in human communication. Archaeologists and historians can observe these long term changes because they adopt a long term perspective which has been termed the *longue durée* (ibid: 27). Braudel has argued that "each 'current event' brings together movements of different origins, of different rhythm" (1980: 34). Therefore a long term perspective allows a better understanding of social processes than the concentration on single events, because it can acknowledge more processes on different temporal scales. On the other hand the retrospective view means that archaeologists and historians do not face innovation with the same uncertainty and unpredictability as economists and politicians (van de Leeuw 1989). They already know the outcome and the path which history has taken and this knowledge reduces the awareness for the adaptive potential available at the time.

2.5 INNOVATION AS A RESPONSE TO CHANGING CIRCUMSTANCES

"Necessity is the mother of invention"

Deciding which action is the most appropriate in a particular situation involves the evaluation of alternatives and results in choosing the option that is seen as leading to the desired outcome with a minimum of effort or cost. This process therefore always includes assumptions about the future. However throughout most of our lives we do not consciously weigh the advantages and disadvantages of different behavioural options but stick to a routine: solutions we chose earlier and that have proven to work. Thus initial conscious decisions can turn into subconscious behaviour (Schimanck 2000: 92-95). Behaviour slipping into our unawareness has the advantage of saving us a lot of time which we can use to do other things.

Stress is perceived if an individual finds itself in a situation for which its habit cannot provide a "custom-made" solution (VanGundy 1987, Winder 2004). Stressful situations, although or because they are unpleasant for the affected individual, can therefore provide fertile ground for innovations by forcing individuals to leave traditional ways of doing things and thinking. Because the unthinkable might become commonplace and the existing order is challenged in times of social or political upheaval, uncertainty provides circumstances in which changes or innovation are more likely to occur within a population (van de Ven 1986, Thurston 1999).

Within the macro economic literature such stress is often associated with competition for or within the market (Metcalfe 1986). Competition is often seen as desirable or even a necessity because it creates incentives for firms and organizations to improve their efficiency. One way of doing this is through innovating. This perspective implies policy-relevant questions of market regulation. This political significance explains the substantial research that has been put into investigating the relationship between market composition, competition and innovation (Ahn 2002).

A significant increase in population density and consequential pressure caused by the need for more adequate or different food supplies has been used as an explanation for cultural and technological changes in archaeology for a long time. Cunliffe even claims it to be "probably the most important single factor bearing on social change" (1991: 525). The 'wave of advance' model developed by Ammerman and Cavalli-Sforza (1971, 1973) argues that the spread of the Neolithic Revolution from the Fertile Crescent through Europe can be explained by an increase in population and modest local migration activity. Population increase is a crucial factor in this model.

Also, external factors like climatic changes are often held responsible for social and technological change. Climate changes often influence the availability of resources. If, consequently, the land no longer provides for a group of people this demands some form of readjustment (Cunliffe 2000: chapter 5). Because old routines might no longer be suitable in changing conditions, they provide a stronger incentive for change or even make it a necessity. This argumentation is based on the assumption that technology can be seen as the means by which humans adapt to their environment (Binford 1965). This view has been criticised because many technologies were applied for purposes different from the ones that became commonplace at a later stage. It has been argued that the mere observation of this later usage might therefore not necessarily reflect why an innovation actually spread in the first place, e.g. the invention of the wheel, which was first introduced for ceremonial purposes (Pfaffenberger 1992). While pressure can have a positive effect on the production of new technological devices, it has also been pointed out that technological innovations are more likely to fail at such times than in times of social security (Fitzhugh 2001).

This shows that the connection between stress and innovation does not have to be entirely positive. However stress is generally seen as an incentive for innovative activities based on the assumption that "a changing, variable environment demands flexible behaviour [...]" (Koestler 1967: 111). However most publications do not consider different levels of competition or stress which might have a crucial influence on these processes. As Koestler (1967) has pointed out there are two possible reactions to a crisis: the emergence of new behaviour to adapt to the new situation (innovation) or alternatively a breakdown of behaviour "when in danger or in doubt, run in circles, scream and shout" (111). While a certain amount of anxiety might stimulate innovation, too much pressure might paralyze individuals and their ability to accept change.

2.6 INNOVATION AS A SOURCE OF RISK

Technological advances and innovations are not always welcome and may be criticized for their negative effects on the quality of human lives as well as the environment (Swearengen & Woodhouse 2001). Like all complex social processes innovations can show indirect or latent effects which might occur with considerable delay and can therefore not be related to the relevant innovation (Rogers & Shoemaker 1971). Research has shown that individuals tend to consider such indirect and long-term effects as less important in their decision processes than effects that show in the near future (Schimanck 2000: 78-79). Long-term effects are therefore more likely to be neglected in cost benefit evaluations. Furthermore, responsibilities for such effects are more difficult to allocate. Critics of technological development often argue with such long-term consequences like environmental pollution or climatological effects. Even though technological developments and innovations are rejected, the conclusions are the same as within the groups that approve of innovation: for different reasons both perspectives want to control and manage innovation.

The critical evaluation brings further aspects into innovation research; firstly questions of responsibility and social change and secondly the question of how innovations and their predicted impact on society should best be assessed. Which time and spatial scale provides the best outlook on the relevant consequences of an innovation? Is it the perspective of a company which might or might not experience an increase in profit? Or the perspective of future generations which have to deal with long term consequences? It is obvious that these questions are of direct concern to politicians. Even though governments have an impact on innovative processes through research funding as well as through policy regulations (Cohen & Levin 1989, Dosi 2000), can politicians still control businesses in times of globalisation?

The risk for the innovator has also been discussed, often in regard to the human reluctance towards change (Hägerstrand 1967: 149-150, van der Ven 1986, Fagerberg 2005). As Schumpeter put it "it is not only objectively more difficult to do something new than what is familiar and tested by experience, but the individual feels reluctance to it and would do so even if the objective difficulties did not exist" (1949: 86).

2.7 RECOGNIZING INNOVATION: ESSENTIAL BUT IMPOSSIBLE?

Schumpeter defined innovation as "any 'doing things differently' in the realm of economic life" (1939: 84); any observable change of behaviour can therefore be seen as an innovation. Economists however favour innovations that produce observable patterns in economic performance, particularly economic growth. Therefore their measurements of innovative activities have concentrated around observable input and output into the market as well as technological change. Considering the importance and vast interest that is accorded to innovation the lack of satisfactory measures of innovative processes, for innovative inputs as well as outputs, should be surprising (Jaffe 1986, Basberg 1987, Cohen & Levin 1989, Acs et al. 1992, BCG 2003, Smith 2005). The economic and political sectors invest a significant amount in innovation, even though it has been admitted that its value and economic return cannot be satisfactorily assessed. Within the economic literature expenditure on R&D or the number of personnel engaged in R&D have been used as a measurement for innovative input while the amount of patents or innovation data have been used to reflect innovative output and technological change (Basberg 1987, Cohen & Levi 1989, Feldman 1994: 29-31, Ahn 2002). However most of these indicators are indirect measures and can only capture certain aspects of the innovation process (Basberg 1987). While patent data has the advantage of being easily obtainable and allows for a long-term perspective, it has been criticized for giving an incomplete picture of innovative activities. Not every invention gets patented and patent statistics cannot differentiate between product and process innovations. Furthermore not all patents are of the same economic value or technological significance to an organization and many are never commercially exploited. At the same time some economically significant innovations are not patentable, like computer software. Because the cost of patents varies within different economic fields these faults may be widened through patent discrepancy between different sectors (Basberg 1987, Cohen & Levin 1989, Griliches 1990, Ahn 2002). This connects to the different temporal perspectives on innovative processes discussed in section 2.4. It is not possible to observe innovation while it actually happen but only in an retrospective.

Also the speed by which an innovation diffuses over time is hard to judge, because there is no means to define the exact moment an innovation is accepted by a unit of adoption (Dosi 2000).

Archaeologists rely on artefacts and the context in which they are found for the reconstruction of human behaviour. Because artefacts often reflect technologies available to a certain group, innovative behaviour examined by archaeologists is often technological (Ottaway 2001, van der Leeuw 1993). Unlike the economic perspective which separates technological from social change, archaeologists claim that technological development cannot be understood without considering the social context in which it took place (van der Leeuw 1993). However such an indirect reconstruction of social processes from static artefacts is problematic (Binford 1983).

Cultural geography studies innovation as visible features in the landscape, which includes the investigation of its spatial origin as well as its spread from the area where it first occurred (Hägerstrand 1967: chapter 1). This implies the assumption that patterns observable in the landscape are caused by human behaviour (Winder 2004), whereby different innovations have different impacts on the landscape. This approach assumes that innovation have an impact on both time and space because they lead to structural changes in the socio-environmental system of a society. These changes can be observed as spatial patterns (Carlstein 1978). Time-Geography tends to be descriptive rather than providing an explanation for the patterns observed. While Hägerstrand (1967: 263-268) assumes that economic as well as psychological factors are held responsible for visible delays in the diffusion process, he does not provide any further proof.

So far no approach has been able to identify innovation while it is actually happening. None of the different perspectives has managed to bound the phenomenon in such a way that it can be unambiguously observed. With abundant causes and consequences that appear on different time and spatial scales, innovative processes seem to be only observable in a retrospective view. This reflects a general problem when dealing with social change and makes it difficult to decide where and how to intervene (Rosenhead & Mingers 2001).

2.8 CONCLUSIONS

Innovations are unambiguously considered to play a crucial role in the course of history, influencing social change as well as technological development, whereas both are closely interlinked. Independently of how such change is evaluated, there is a strong desire to intervene in innovative processes. Although all disciplines identify knowledge as one of the major sources of innovation, its role is considered from significantly different angles (see figure 4). The political and economic literature focuses on knowledge as 'technological know-how' and generally ignores its cultural implications. It is felt that such a biased view is insufficient, considering the complexity of innovative processes and their impact on various scales and parts of society. Even if the sole interest lies in explaining purely economic processes, only a wider concept of knowledge can provide a valid explanation.

The understanding in Time Geography and archaeology is more comprehensive as it includes cultural and individual aspects as well as technological know-how. This study will therefore focus on the broader understanding of knowledge used by social geographers, archaeologists and sociologists. However its application to the archaeological record will meet problems in regard to this factor, not because of the archaeological perception of knowledge but due to the nature of the archaeological evidence. Compared to economists or EU-politicians, archaeologists have only very limited access to human beliefs and other cultural aspects.

The macro economic literature focuses on – relatively short term - effects on market and technology development. Even though this approach points towards future developments and their successful management it ignores consequences that go beyond economics or that show with considerable delay (Dosi 2000). The retrospective view of archaeologists on the other hand allows the observation of such long term consequences and has led to the identification of different factors considered crucial in innovative processes.

One important aspect of innovation is that it is a process rather than a 'thing', which makes it difficult to observe and to pin down. "it is a serious mistake to treat innovation as if it were well-defined, homogeneous thing that could be identified as entering the economy at a precise date – or becoming available at a precise point of time [...] The fact is that most important innovations go through drastic changes in their lifetime" (Kline & Rosenberg 1986: 283).

The importance associated with innovations for all areas of society explains the strong desire to control and manage such processes. However, despite abundant research, none of the disciplines is yet able to even measure or observe innovation satisfactory.

This chapter has shown how different perceptions of innovation are even in the disciplines of the social sciences. That is because innovation is a complex social phenomenon and thus a 'wicked' problem. Reality and value judgements vary significantly between and even within one discipline (Fagerberg 2005). Economists consider technology development as crucial and thus bound innovation as technology development; correspondingly the solution for the desired increase in innovative activities is to invest in R&D. The different research interests clearly influences the bounding of the problem. In order to talk about innovation a conceptual model is needed.

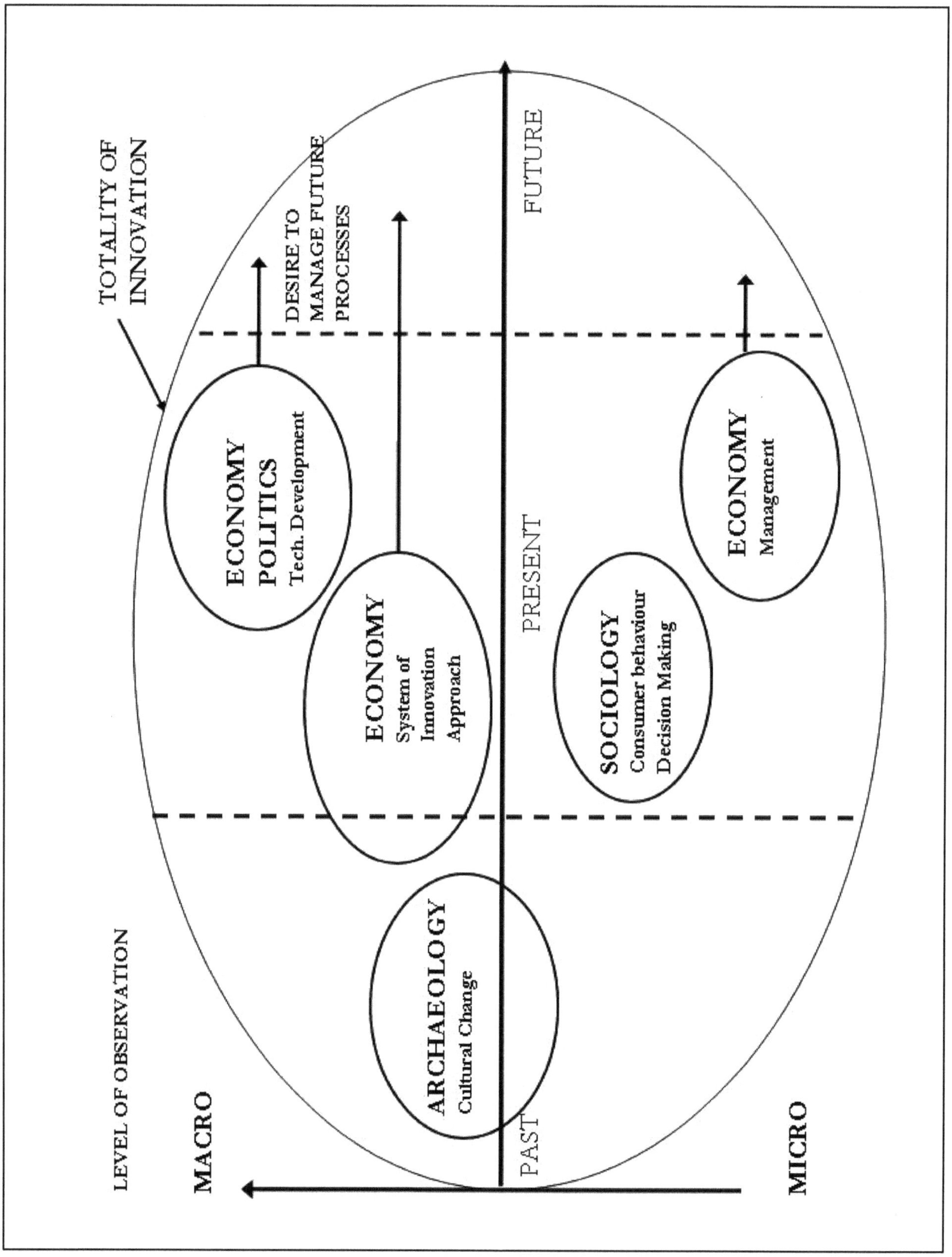

Figure 4: *Variety of definitions on innovation*

3. THE SOCIOLOGICAL CONTEXT

The literature review on innovation has unambiguously identified 'knowledge' as a crucial component in the process of innovation. The subject, however, is considered from two very different angles: as technological know-how (the economic and political perspective) or as the reconceptualisation of cognitive structures (archaeology and cultural geography). This study will focus on the latter and defines knowledge as the shared beliefs of a social group. All knowledge is socially constructed and changes over time. This concerns values, norms as well as our understanding of the 'truth'. This chapter will justify this assumption by summarizing the sociological understanding of how humans interact with others and their environment. It will demonstrate that the human constitution requires shared knowledge as a basis for action.

Sociologists assume that human interaction contains behaviour (action as well as omission) which, accumulated to collective action, can alter or reinforce social structures (Giddens 1984, Archer 2005). These structures will influence future actions by reducing or increasing the option space of individuals. The human condition is thus determined by two processes: a top-down (structures enabling or constraining human behaviour) and a bottom-up (collective action that can transform or sustain structures) process (see figure 5). Only collective action can result in social structures, the action of one individual alone can never generate or alter structure (Schimanck 2000: 12). However sometimes a single event or action can have a visible impact on the macro level of society, because it may generate a pattern of unintended but extensive consequences. One example is the assassination of Franz Ferdinand which resulted in the First World War (Giddens 1984: 13).

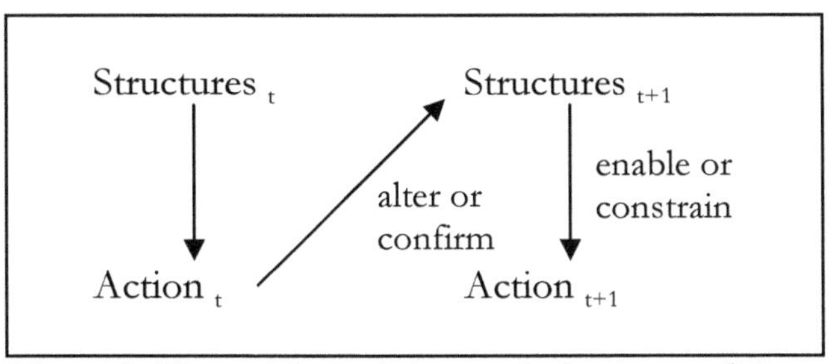

***Figure 5**: The sociological view on the human condition*

The crucial aspect of this scenario is the relationship between the micro (individual action) and the macro level (social structures). Both are closely intertwined but by no means circular as "structure logically predates the action(s) which transform it and structural elaboration logically postdates those actions" (Archer 1982: 468). This means that actors are influenced by structures that they did not create themselves. Structure and action operate over different time intervals and, although collective action might transform or reproduce social structures, actors do not actually create structures, they merely remake what was already there (Giddens 1984: 160 – 174, Archer 1982, Read 2003).

Sociology has created an artificial separation between these two processes. Up to the 60s the discipline was dominated by approaches concerned with the impact of structures which were then replaced by theories of interaction. Attempts to combine both processes have only recently appeared in the work of Giddens, Bourdieu or Archer (reviewed by Hradil 1992). The following section will give an overview of the understanding of action and social structure and their connection.

3.1 SOCIAL STRUCTURES

"Culture provides for the members of a society a conceptual universe that both frames and constructs patterns of behaviour" (Read 2003: 17)

Humans have a great ability to learn and adapt. During socialisation individuals become accustomed to existing structures which express themselves through values, norms and within the *Weltanschauung* (world view) of every social group. They thus learn to act and think in a socially accepted and expected way. Socialisation influences the way we make sense of our surroundings but also reduces the option space of our actions by defining the unthinkable and undoable. It is crucial to point out that social structures are culturally constructed and thus highly dependent on human consensus, even though the individual itself might experience them as an objective universal truth which exists outside human agreement (Schimanck 2000, Bourdieu 1983, Garfinkel 1973).

The focus on structures as a means of adapting an individual to society has a long tradition within sociology. Emile Durkheim (1950) tried to establish sociology as an independent science at the end of the 19th century and defined *'social facts'* as its unique research objects. A social fact is "every way of acting, fixed or not, capable of exercising on the individual an external constraint" (1950: 13). Durkheim described them as the 'casting moulds' of society; controls of conduct which imprint socially accepted actions and beliefs through the process of socialisation and education. He considered social facts to be external to the individual and "endowed with coercive power, by [...] which they impose themselves upon him, independent of his individual will" (1950: 2). Through this force social order is maintained as individual interests are controlled and subordinated to the interest of society.

Parson's (1973) *'theory of action'* claims that individuals follow the expectations of the social system rather than their own intentions. Every human has various social roles, they are a parent, a teacher, a manager etc., and these roles are attached to normative rules of action: a manager is expected to behave differently from a parent for instance. Humans incorporate these expectations. Sanctions assure the obedience of these norms and because humans have a tendency to avoid punishment they follow these rules. Through this mechanism society and individuals become compatible.

Bourdieu's (1983, 1986) work focused on how social classes reproduced themselves. He studied the education system in France and developed the concept of *habitus* as a connection between structure and practice (human action). Habitus is defined as a disposition towards the world. It becomes embodied through socialisation and plays a crucial part in social reproduction. Through the process of embodiment the individual becomes at 'ease' with the behaviours and beliefs of its social milieu. Breaking these behavioural patterns is difficult and sometimes impossible because its incorporation reduces the ability to adopt possible alternatives. Bourdieu thus claims that there is a correlation between social position and habitus. Privileged children acquire the lifestyle expected in the educational system and expressed by teachers and thus naturally fit into the world of educational expectations, while children from unprivileged backgrounds do not. Bourdieu has shown that despite an apparent freedom of choice in arts, people's preferences strongly correlate with their social position. The same is true for aspects of language, like accent, grammar etc. Through this mechanism social capital[1] gets transferred from one generation to the next and social stratification is reproduced.

All of these three approaches describe how individual action is influenced by the existing structures of a group. This influence is often indirect and structures act as mental filters or cognitive patterns and thus have an impact on our actions. The information humans perceive from their surroundings is too abundant to act or decide upon. Cognitive patterns adapted through socialisation and education allow us to reduce complexity and attach meaning in a way that enables us to make sensible decisions. As with behaviour, social structures increase or reduce the space of possible cognitive patterns. The values, rules, morals and expectations an individual incorporates and uses as a basis for its decisions and actions are therefore strongly influenced by the social structures valid within the group it grows up in. Once incorporated, categories and cognitive patterns can be applied to future situations and will thus help to classify and explain the unknown. Humans cannot tolerate meaninglessness and through categorising information with commonly accepted patterns they substantially decrease uncertainty and their actions and thoughts become predictable for others (Willems 1997: 59).

Schimanck (2000) differentiates between three different kinds of social structure, structures of expectation ("*Erwartungsstrukturen*"), structures of interpretation ("*Deutungsstrukturen*") and constellation structures ("*Konstellationsstrukturen*"). Many of the structures included in the first group manifest themselves through the expectations attached to social roles, like norms, morals, customs etc.; they also include rules and laws. Many of the structures discussed above are included in this group. Constellation structures describe interactions between actors that are fixed in such a way that they follow the Nash equilibrium, so that none of the actors involved - following his own intentions - could have acted differently to achieve a more desirable outcome (Schimanck 2000: 173-180). Structures of interpretation determine the cognitive and evaluative orientation of individuals and thus have an impact on their thinking. Like all other structures, structures of interpretation contain categories which humans apply to the world in order to reduce complexity. Influenced by our beliefs system, newly selected data are related to already existing categories and thus put into a wider context. Therefore everything that comes to our awareness is the result of a selective process, strongly influenced by our individual values and beliefs. Searle talks about a "background of know-how that enables me to cope with the world" (1998: 108) and contains a vast apparatus of fundamental knowledge which underlies all our intentional states and is prior to all our beliefs. Archer has described this kind of embodied knowledge as second nature; "It is a 'knowing how' when doing, rather than a 'knowing that' in thought" (2000: 162).

Unlike other expressions of social structure, beliefs are experienced as part of oneself rather than the socially constructed artefact they actually are. While some structures, for instance expectations attached to social roles, regulate what an individual *should* want to do, beliefs actually constitute its intentions. While structures of expectation might often collide with an individual's intentions and are often perceived as an "annoying fact" (ärgerliche Tatsache) (Schimanck 2000: 67-68) beliefs are considered as a crucial part of oneself.

Human are mostly unaware of the way they conceptualise the world. As Koestler has described it "we are obeying the rules without being able to define them" (1967: 108). All the information that humans perceive is inevitably interlinked with the beliefs according to which they filter and categorise these so-called facts (Vickers 1983). As cognitive patterns become incorporated they slip into our unconsciousness and it is only when confronted with behaviour based on beliefs very different from our own that we become aware of this mechanism. People with different experiences, cultural backgrounds or information might *frame* the same situation very differently (Willems 1997, Bourdieu 1983) and their perceptions and reactions might be very different. It is often only through an encounter with such individuals that we become aware of the cultural rules and constraints that influence our actions.

[1] Bourdieu (1986) widens Marx' concept of capital and economic theory by introducing three types of capital: economic, social and cultural. *Social capital* includes resources based on group membership and networks of influence and support while the concept of *cultural capital* describes an individual's skills, forms of knowledge and education. The latter expresses itself in an embodied form as cultural habitus. While economic and social capital might change (one can win the lottery or gain access to a new group of people) it is very difficult to change habitus and overcome these embodied patterns.

Humans are able to change or widen their belief categories over the course of a lifetime - young children do this all the time as part of their socialisation - however, beliefs that build the basis for our actions as well as our understanding of the world are often too "precious" to alter. Beliefs systems can theoretically be separated into two groups: essential and inessential. While the latter include categories for classifying objects or behaviours that are possible but not necessary, very loose values and likings, the first category contains beliefs that define the foundations of our thinking and acting and determine our intentions (Winder 2005: 50-51). However inessential beliefs are still mental artefacts, rather than pure facts, since they have been accumulated through the same process of culturally constructed perception (Vickers 1983: 348). Humans might be unaware of gradual changes in their inessential beliefs, but they will always recognise challenges to their essential beliefs. Such a confrontation can be a very stressful experience and is likely to trigger an emotional response and rejection. It also brings the potential of recognising and accepting previously unseen alternatives and has therefore been described as an 'epiphany' for the affected individual (Winder 2005: 83-85).

The concept of interacting structures, perceptions and behaviours is not only found in sociology. Vickers' (1995: chapter 4) concept of *'appreciative systems'* brings all these factors together in his model of decision-making. Human action, communication and application of meaning are all influenced by appreciative systems. Every decision-making process is based on three kinds of judgement; by observing 'what is' (reality judgement) and comparing it with 'what should be' (value judgement) an individual evaluates whether, and which, actions should be taken (operational judgements). Value judgements represent an individual's desires and intentions and follow a *world-to-mind direction*, reality judgements define how things are and follow a *mind-to-world direction* (Searle 1998: 100-103). If our desires fail to fit the mind-to-world direction, it is not because they are wrong but because the world is not how we want it to be.

Our perception of reality and our decision-making processes are highly influenced by our tacit norms or values. Value judgements are the result of individual experience and influence every perception of reality and, at the same time, limit the ways in which humans organise their experiences or perceive their surroundings. "The development of an appreciative system, at once enabling and limiting, is the inner history of an individual, an organisation and a society" (Vickers 1983: 69). Our values might be changed during a decision-making process, in which case the next appreciative cycle is based on different conditions. "An appreciative system is a process whose products – cultural manifestations – condition the process itself" (Checkland & Casar 1986: 5).

Appreciative systems are also used by units larger than the individual; organisational decisions for instance are based on the same mechanism (Vickers 1983, Checkland & Holwell 1998). In reality many different appreciative systems operate simultaneously both on different time scales and levels of social organisation (Checkland & Holwell 1998).

3.2 VERSTEHEN: ACTION AND INTENTION

Due to the theoretical dualism between structure and action theory, sociology often separates human behaviour from structural influences. Bourdieu is one of the few sociologists who aimed at overcoming this polarisation. However in practice his theory of culture tends to neglect the aspect of structure generation (Hradil 1992).

While an action itself can be obtained quite easily through observation, the underlying intentions and values that constitute this action are much harder to acquire. This however is crucial to the understanding of human action. One of the founders of sociology, Max Weber (1978) declared that gaining an understanding of the subjective meaning (*subjektiver Sinn*) behind an individual action should be the core research interest of sociology. It is this understanding rather than the actual action that is crucial for social processes. Subjective meaning can be obvious or inward and personal.

Weber (1978) uses subjective meaning to oppose action (Handeln) to behaviour or pure reaction (Verhalten). Anthony Giddens, on the other hand, has argued that sociology should "separate out the question of what an agent 'does' from what is 'intended' or the intentional aspect of what is done. Agency refers to doing" (1984: 10). This does not contradict Weber's concept of 'subjective meaning' but shows the different research interests of both sociologists. Giddens wanted to overcome the dualism between structure and action and was therefore interested in the entire action spectrum exhibited by humans, regardless of whether it is indented or not. This includes all forms of human action, routines and habitus as much as rational choices. Weber on the other hand wanted to understand human action rather than explaining the constitution of social structures.

The relationship between an individual's intention and behaviour is by no means unambiguous. Two people acting in the same way may have very different reasons for doing so while at the same time equal intentions may express themselves in different actions. The German sociologist Georg Simmel (1908) describes every social phenomenon as a dichotomy, composed of the two elements, *form* and *content*. While the content of an interaction holds its purpose or motive, its form describes the mode of interaction through which a specific content becomes social reality. Murder, for instance, can be the form of the two very different contents of self defence or revenge. Simmel states that this separation can only be drawn analytically and both are inseparable in reality. Searle (1998: 99) makes a similar distinction between the *type* and the *content* of intentional states[2] "[...] for example, we can hope that it will rain, fear that it will rain, and believe that it will rain. In each case, we have the same content – that it will rain – but the content is presented in different intentional modes".

For the majority of the time humans do not act within a vacuum, which means that they have to consider other actors in their choice of action. Weber (1978) defined action that is aimed at other people as 'social action'. Sociologists always study actors in constellations with other individuals when their option space is influenced by the behaviour (omission or action) of others. The goals and intentions of different actors often interfere and actors can either dissolve the constellation or try to find solution to overcome these interferences (Schimanck 2000: 173- 180). Read (2003) has used the term *adjudication* to describe attempts to resolve conflicts that "inevitably occur between behaviours arising out of one's identity versus behaviour arising out of one's assessment of the consequences of action". This adjudication can take place on the level of the individual as well as on the level of social norms or laws

Three models of human action: homo sociologicus

In order to operationalise meaningful action and its underlying intentions Weber (1978) defines four ideal types of action (affectual, value-rational *"wertrational"* , traditional and instrumental-rational *"zweckrational"*). These ideal types broadly overlap with the three main sociological models for human action: normative, rational and emotional man. (see figure 6)

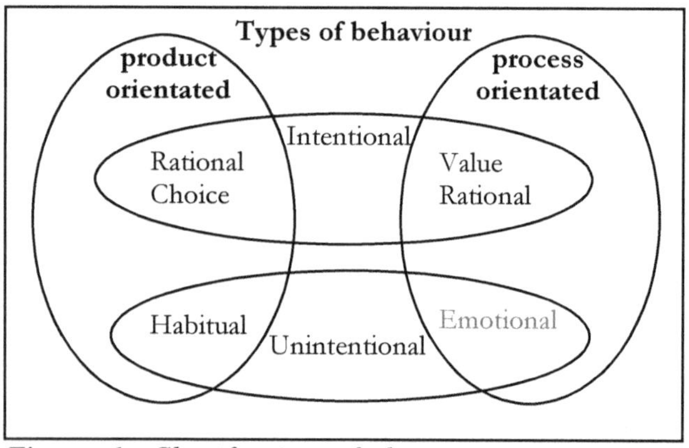

Figure 6: *Classification of the motivation initiating different kinds of behaviour*

[2] Searle uses the term *intentional* to describe the "special way the mind has of relating us to the world" (1998: 100). It is important to note that this does not necessarily imply 'intending' and that not all intentional states are conscious. Searle's definition therefore does not coincide with Weber's division of behaviour and meaningful action. His intentionality includes conscious and unconscious behaviour and thus covers a greater range than Weber. In order to understand his definition of intentional state we have to ask under which conditions an individual's desire would be satisfied.

The *'normative'* model or homo sociologicus is the traditional paradigm in sociology and played a crucial part in the development of the discipline. It states that human behaviour is formed by the norms of society, expressed through role expectations. Many of the sociologists introduced above fall into this paradigm: much of Durkheim's work has described how action is formed by social structures and Parsons (1937) has argued that actors follow norms attached to social roles. Bourdieu's (1983) concept of 'habitus' and Weber's *traditional* action also fit into this category. Of course a sole focus on norm acceptance cannot provide a complete explanation of how human action and decisions are formed and this model has triggered a lot of criticism (Goffman 1974). It has been argued that human behaviour is as much passive 'role-taking' as it is active 'role-making', because it is often influenced by personal desires and wishes rather than social expectations. Role expectations can contradict individual wishes and, because humans always take up more than one role, the expectations attached to them might contradict each other and create a series of intra and inter role conflicts. Humans have the ability to circumvent existing norms, as Koestler puts it "[...] the rules permit a certain variety of alternative choices" (1967: 105), structures and expectations always allow the actor a certain degree of freedom. Where there is no choice at all, the behaviour turns into a specialized reflex and becomes deterministic.

Normative behaviour can easily become repetitive in which case the individual merely follows solutions it chose earlier and that have proven to work. Most human behaviour takes the form of such routines (Weber 1978). This kind of behaviour is likely to appear within a stable social environment with clear role expectations or effective social control, both external and internal (Schimanck 2000: chapter 6).

Rational man

The second model of action, the 'rational' paradigm was developed in classical and neo-classical economics. It regards action as determined by goals and a rational evaluation of the different means of achieving these goals. Weber's definition of *instrumentally-rational* action is in accordance with the ideas of the rational choice approach or the so-called 'homo œconomicus' (Becker 1981, Esser 1990). This type of action is characterised through rational calculation, aimed at achieving a chosen goal following the principle of cost minimisation. These cost benefit calculations initially require conscious decisions but can quickly turn into habitus if the evaluation is not repeated and the individual sticks to the behaviour even though the circumstances that determined the value of this action might have changed.

While the instrumentally-rational actor focuses his action on a goal, the *value-rational action* is defined through the importance attached to the process of acting rather than goal achievement. This dichotomy goes back to Aristotle and his definition of 'praxis' where the meaning lies within the act and 'poiesis' where the intended meaning can only be achieved if the individual succeeds in achieving his goal. Vickers argues very strongly that it is processes (relationship maintaining) rather than goal achievement that motivate human actions and give meaning to life. "To get the job or marry the girl is indifferently an end, a means and a goal; it is an opportunity for a new relationship. Though the object of the exercise is *to do* the job and *live* with the girl; to sustain through time a relationship which needs no further justification, because it is or is expected to be satisfying in itself" (Vickers 1972: 128).

Checkland's definition of purpose as "an end which can be pursued but never finally achieved (as can an objective or goal)" (1998: 316) supports the idea of processes rather than goals as determinant of human intentions. While conscious human action always includes both a goal and a process, actors put stronger emphasis on one of these two categories depending on their values. In disparity with the rational choice approach it is argued that conscious action is aimed at processes rather than goals in the majority of the time. Both types of behaviour in their pure form are predictable.

Emotional man

The third paradigm of human action, the concept of 'emotional man' was introduced by Flam (1990) to complement – not replace – the normative and rational approach. An important aspect of this model is that emotions are unique in their unpredictability. "Feelings have involuntary character. They cannot be produced by will. Feelings invade or overwhelm. They connect or separate individuals against their will" (1990: 43). This kind of behaviour has been classified as *affectual* by Weber.

Weber had to acknowledge himself that his categories are not always logically coherent with his own definition of meaningful action. Emotional action for instance often takes the form of an uncontrolled reaction to some stimulus and the majority of traditional behaviour exists as an almost automatic reaction which has been habitually accustomed (1978). Both types of action can therefore not be strictly classified as meaningfully orientated. Even though Weber's classification does not allow for a strict separation between conscious and subconscious or intended and unintended action it is felt that this general differentiation is a very useful one in regard to innovation.

Checkland (1993, 1998) uses the expression *purposeful* to describe intentional action based on selective perception of the world. He describes purposeful action as deliberate, decided and willed and thus differentiates it from "habitus". In his application of soft system methodology to 'real-world' organisations and management decisions he focuses exclusively on this form of human behaviour. This imposes the presumption that all management decisions are performed after a conscious evaluation of perceived data based on existing values, beliefs and professional knowledge (Checkland & Holwell 1998). However Checkland does not provide the methodology to test this assumption. As the intention underlying an action is hidden within a person's mind its classification as routine or willed cannot be obtained immediately but requires observation over a significant time span.

Another important term introduced in Weber's typology is "value" (1978). Human action often seems to be attached to values. A rational choice always includes individual decisions about which goal is desired most and which ways of achieving it are considered acceptable, both of which depend on a subjective evaluation by the individual. While the advantage of habitual or normative behaviour lies within the fact that such an evaluation is unnecessary, the value attached to this category lies within the significance the acting (again a process) has for the individual. This meaning might only become apparent when the routine is taken away. An individual therefore need not be aware of the value attached to its actions. Figure 4.3 shows a combination of several of these categories to classify human action. The classification of process and product orientation is suggested in combination with intentional and unintentional behaviour.

3.3 CONCLUSIONS

This chapter has set up the background to the model of innovation. It has described the two processes that constitute the human condition: social structures influence the option space of each individual and thus their choice of action. These processes act from the top down. Collective human action can transform or confirm these structures from the bottom up. Human behaviour is never fully determined by social structures (humans have the ability to breach norms or innovate) and collective action often results in emergent structures and unforeseen or unwanted consequences. This makes the model of the human condition dynamic and social change nonlinear.

The chapter has also demonstrated how human interactions involve socially-constructed knowledge. Humans are incapable of separating themselves from their embodied beliefs and mental filters. Everything they do is influenced by their value systems and former experiences. Considering the various definitions of innovation discussed in the previous chapter, the author argues that it is this embodied, tacit knowledge rather than technological know-how that should be considered in regard to innovation. This does not imply that technology is not seen as crucial for at least some innovations but rather that it is not part of the basic mechanism. Even seemingly objective scientific truths (the earth is the centre of the universe or humankind has been created by God) have been filtered by the appreciative systems of individual scientist and scientific peer groups and can therefore change over time.

4. A CONCEPTUAL MODEL OF INNOVATION

Humans have a great ability to learn, rather than purely react to stimuli, because during the course of evolution this has proved a powerful adaptation to a changing environment and rapidly shifting cultural demands. As we have seen in chapter 2 the literature is unified in acknowledging the importance of innovation as a mechanism of change. Innovations are generally evaluated as good and desirable and a significant amount of money and effort is invested in order to stimulate them. Despite all these investments, innovation appears to be relatively rare. If learning to deal with the surrounding world is part of human nature, why is there such a great amount of non-reflected behaviour in our daily lives? Why do humans often seem to have a blind spot when it comes to choosing their behaviours, consciously or not, making them unaware of all the available opportunities in a particular situation? The answer to these questions is closely related to the human condition and introduces another important aspect of innovation; its ability to push our knowledge about the world past its sell-by date creates insecurity and social exclusion.

4.1 WHY PEOPLE INNOVATE SO LITTLE

"*... there is no more delicate matter to take in hand, nor more dangerous to conduct, nor more doubtful in its success, than to step up as a leader in the introduction of change*" Machiavelli

There are basically two major advantages in responding to life in a routinised and non-reflective way: on the one hand it conforms to the principle of parsimony and saves time (Koestler 1967, Bourdieu 1988, Schimanck 2000, Weber 1978) and on the other hand it allows us to predict the behaviour of others and protects us from social exclusion (Goffman 1974, Willems 1997).

In general, human decisions are based on an evaluation of possible alternatives resulting in the choice that promises the desired outcome with a minimum of effort and cost. This process always includes assumptions about the future. The factors involved in this decision-process do not have to be monetary and can include aspects like fear, affinity, avoidance or gaining of group acceptance. Every cost-benefit calculation is based on an individual's personal perceptions, values and experiences. The behaviour chosen by people in similar situations can therefore differ significantly. But every estimation does require time and information and both are valuable and expensive. Furthermore humans only have limited abilities for information processing (Janis & Mann 1977: 21-33). Therefore they often do not search for the perfect solution but stop evaluating as soon as they find an acceptable way of achieving their goal. This is especially true for so-called 'low cost' situations, where the desired outcome is not considered particularly important by the acting individual, e.g. deciding which washing powder to buy or which socks to wear. It is only for rare 'high cost' decisions, when the result of the chosen action is believed to have a great impact on the individual's future, that humans will invest a lot of time and effort into making the right decision. They are 'satisfiers' rather than 'optimizers' (March & Simon 1958, Hägerstrand 1988, Esser 1990, Schimanck 2000: chapter 4). Throughout most of our lives we do not consciously weigh the advantages and disadvantages of different behaviours but rather stick to a routine. Often such routines are based on earlier choices that have proven to work. Thus initial conscious decisions can turn into subconscious behaviour. Behaviour slipping into our unawareness has the advantage of saving us a lot of time which we can use to do other things. We would not get very far in our complex world if we had to start every day by making conscious decisions about whether it is less time-consuming to make coffee before taking a shower or the other way around.

Although routines enable us to concentrate on other things they also lower our awareness for alternatives and we often do not even check whether our behaviour is still appropriate. For instance we might take the car to work every morning, not realizing that the conditions determining our initial decision have changed and that the lack of parking spaces would now make the bus far more convenient (Schimanck 2000: 92-95). Van de Ven (1986) has opposed routinised to innovative behaviour. Interested in organisational innovation, he argues that while individuals gradually adapt to their working environment they become unaware of factors that could improve their work performance and this mechanism is likely to decrease the amount of innovative behaviour. The advantages of routine behaviours prevent them from spotting situations where changes could improve their lives. Humans are less likely to perceive very gradual change which reduces the likelihood of change in slowly changing environments (van de Ven 1986). Unawareness becomes habitual and routines inhibit innovative behaviour. Consequently the advantages of routine and habitus are also their disadvantages "the very nature of fixed habits of thinking, their energy-saving function, is founded upon the fact that they have become subconscious [...] are proved against criticism and even against contradiction by individual facts. But precisely because of this they become dragchains when they have outlived their usefulness" (Schumpeter 1949: 86).

The second advantage of routines and habitus becomes apparent when we look at human interaction in a complex world. While humans have a certain freedom of choice within existing boundaries they are constrained by the decisions of other people; the car might be the best way to get to work unless everybody else decides to do the same and there is a major traffic jam. Consequently humans have an interest in observing and controlling the behaviour of the people surrounding them, so they can use this information as a basis for their own behaviour. We cannot know what other people are thinking and how they are going to act, and as a means of approximation assume that, in general, others are the same as us and will therefore act similarly. Humans have the ability to be empathic (Schimanck 2000: chapter 3). By projecting our decision-making processes and beliefs onto other people it becomes possible to predict how they are going to act. This strategy, of course, is not infallible but it is the best we can do (Goffman 1974). Parsons (1973) has argued that an important aspect of social roles and norms is in allowing us to make assumptions about how other people will behave. Goffman refers to this type of norms as frameworks. "Social frameworks provide background understanding for events that incorporate the will, aim, and controlling effort of an intelligence [...] human being" (1974: 22). Frameworks define a situation and the behavioural norms that go with it.

People who do not follow such role expectations might be experienced as unpredictable or even dangerous, a risk to be avoided. Because humans are highly dependant on others for confirmation of their identity and self, they have a profound interest in staying within the social norms of their group (Goffman 1974, Watzlawick et al., 1974).

Different disciplines have shown that individuals have a tendency to conform to others in their group. The dual inheritance approach has argued that behaviour underlies a biased selection, because individuals tend to copy the cultural traits that are most frequent in their group. This is called the 'conformist transmission' (Boyd & Richerson 1985, Henrich & Boyd 1998). Experiments by Asch (1958) have demonstrated the role of peer pressure in human perception and behaviour very clearly. He placed one person in a group with seven others and asked them to match the length of one given line with one of three other lines, all of different length. The results were announced publicly. The test person was unaware of the fact that all the others were instructed to give a wrong answer. Even though their answers clearly contradicted the senses of the test person, in about a third of the cases he or she chose to conform with public opinion and also matched the wrong line. Some said afterwards that they assumed that the rest of the group was correct and their own perceptions wrong. Others said they knew the answer was wrong but did not want to disagree with the group. In a control group in which members were asked to write their answer down no wrong answers were given. This demonstrates that unwanted deviation from the group may put significant pressure on an individual.

Innovative behaviour is, by definition, a deviation from the norm and breaks away from routines. This bears risks for the innovator as it might put him in disagreement with his group. Innovators run the risk of being perceived as offenders as well as having to act against the human tendency to conform. They cannot expect others to be readily receptive to the new perspective they suggest and might encounter reactions (Schumpeter 1939: 86- 87).

Koestler defines innovation as a successful escape "from the bondage of mental habits" (1967: 178) and acknowledges that in order to do so "one has to overcome immensely powerful intellectual and emotional obstacles" (179). This is due to the fact that mental habits are culturally embedded and provide the basis for our thinking and acting. Before one can adapt to new ways of seeing the world, one has to let go of seemingly self-evident axioms and "to unlearn is more difficult than to learn" (Koestler 1967: 179). Koestler therefore divides revolutionary change into two phases: a destructive and a constructive one, an undoing and re-doing, with the first being the more difficult. Winder has described a change in beliefs as an *innovation jolt* and thus accounts for the impact such a change may have on the individual.

Most of the literature on innovation makes us believe that innovation is a 'good thing'. Economists and Politicians equate it with economic growth and consider it an essential advantage in market competition. "It is almost universally accepted that technological change and other kinds of innovations are the *most important sources of productivity growth* and *increased material welfare* – and that this has been so for centuries" (Edquist 2000: 3). West and Farr define innovation as "the intentional introduction and application within a role, group or organization of ideas, processes, products or procedures, new to the relevant unit of adoption, designed to *significantly benefit the individual, the group, organization or wider society*" (1990: 9). Despite their general definition as something 'desirable' or 'good', innovations may actually bear risks for the individual innovator because they constitute a deviation from secure and tested ways which, besides being difficult in their own right, might also provoke rejection from others.

4.2 THE CONCEPTUAL MODEL

The literature review on innovation has identified knowledge as the most important factor in innovative processes. The previous chapter has shown how knowledge in the form of embodied beliefs and mental filters provides the basis of every human activity. This supports approaches to innovation which define knowledge as shared beliefs rather than technological know-how. Finally the beginning of this chapter has demonstrated how human nature brings the aspect of risk and uncertainty into innovation. A conceptual model of innovation has to account for all these aspects.

This book will adapt Winder's definition of innovation as "a change of beliefs leading to a change of behaviour" (2005: 278) (see box 1) which is considered to be the best representation of all the crucial aspects identified above. His definition splits innovation into two parts; an invisible cause (a change of beliefs triggered by information flow) and an 'outward' or observable effect (a change of behaviour). Both aspects have to be connected but do not necessarily follow each other immediately.

The term 'beliefs' describes the internalized assumptions that provide the basis of every human activity. They can take the form of embodied culture (habitus), tacit knowledge or mental filters and determine the way we evaluate and see our outside world. A challenge to these beliefs can lead to a new perception of our surroundings. The individual can see new opportunities or threats and if it decides to act on this new knowledge and change his or her behaviour we have an innovation.

The suggested definition for the cause of innovation classifies knowledge as 'shared beliefs' rather than information or 'know-how'. It excludes information that merely adds new data without challenging our perception of the world. This view differs significantly from the assumptions found in the economic and political literature (see chapter 2). These disciplines generally define knowledge as technological know-how. Knowledge generation and R&D aim at stimulating innovation by producing more information but not necessarily information that is *qualitatively different* from what we already know. This is a crucial difference and this study makes a strong case for an understanding of knowledge that goes beyond a pure technological definition and relates to the ways in which humans make sense of their environment. It thus challenges the assumed causal relationship of research, knowledge generation, innovation and economic growth (see chapter 2). Others have argued similarly. Hägerstrand for instance stated that in regard to innovation "the sheer mass of information tends to cause congestion" (1988: 226). Humans are limited in their ability to process information and too much information will either be ignored or might even distract people from identifying the actual facts. This argument implies that it is a *qualitative change* rather than quantitative accumulation that is crucial in regard to innovation.

The one-sided understanding of knowledge as technological know-how in the economic literature is considered responsible for some of the difficulties experienced in technology transfer and the commercial exploitation of research results (HoL 1997, Seaton & Cordey-Hayes 1993, Tomes 2003, BCG 2003). It is the human reluctance to give up familiar patterns of thought that restricts the possibility of innovation rather than a lack of information (Hägerstrand 1988).

Although the background for this conceptual model has been discussed on a very general level, the adopted definition tends to support the view of the 'softer' sciences rather than the perspectives found in economics and engineering. By introducing a challenge of beliefs as the cause of innovation it stresses the role of individual and cultural factors that tend to be overlooked within the economic disciplines. Earlier chapters should have made it clear that the concept of 'knowledge generation' is too general to explain innovative behaviour. The model presented here may provide useful insights for these disciplines. It should also meet little rejection from the archaeological or geographical perspective as these disciplines have always argued against know-how as an exclusive stimulant of innovation. An innovation can only diffuse when people accept it, a process also influenced by factors other than know-how or pure cost-benefit calculations (Renfrew 1986, Lemonnier 1993, van der Leeuw 1993, Hill 2002b).

The conceptual model defines the starting point of every innovation at the individual or micro level. Even information that is aimed at an institution, for example, is always received by an individual. The same input might mean very different things to different recipients. In particular the same information might challenge one person's beliefs but not another's.

Furthermore a change of beliefs is a very personal experience. While they are invisible from the outside, they are likely to leave a significant impression on us because they question matters we have taken for granted. The term 'innovation jolt' describes the force such changes may have (Winder 2005:83).

It is important to point out that not every change in beliefs results in a change of behaviour and not every change of behaviour is triggered by a change of beliefs. People can easily change their behaviour without changing the way they look at the world. Both cases describe other forms of social change but not innovation. This research is not interested in changes that do not include reconceptualisation.

In order for an innovation to become visible at the macro level the change of behaviour has to become a collective one. If somebody tells me about a sandwich shop around the corner from my office (information flow that make me recognise a new opportunity) and I decide to buy lunch there from now on (change of behaviour as a response to the new information), it is an innovation. But until other people start to buy their sandwiches there as well it will go unnoticed on an aggregated scale. As discussed in chapter 4, the behaviour of one individual is unlikely to have an impact on the course of history. The conceptual model suggested here therefore includes many small-scale innovations while the majority of disciplines that study innovation are only interested in change observable at the macro level; economists and politicians focus on economic growth and the limited data of archaeologists only allows them to observe large-scale changes. This does not influence the usefulness of the model. It simply argues that the underlying mechanism is the same for every innovation independent of how far it diffuses. Even revolutionary changes begin with the challenge of an individual's beliefs. The relationship between the micro and macro level and the diffusion of innovation will be discussed in chapter 7.

The conceptual model includes new ideas (change of beliefs) as well as the application of this new knowledge (change of behaviour). Transferred to the literature review it combines the terms 'invention' and 'innovation' in a way that makes the separation of the two superfluous. There is no innovation without a challenge of beliefs. This might appear to be of minor importance but it decreases the complexity and ambiguity of innovation terminology.

CHANGE OF BELIEFS	**INNOVATION** → that leads to a	CHANGE OF BEHAVIOUR
internalized presumptions that build the basis of action they determine the way we perceive our outside world		
CAUSE		EFFECT
allows individuals to recognise a new opportunity or threat		appropriate response to this new knowledge

Box 1: *The conceptual model of innovation. Innovation is understood as a process of reconceptualisation.*

4.3 CONCLUSIONS

This chapter has suggested that humans tend to be reluctant to change and that the majority of human behaviour is based on routine and habitus, the exact opposite of innovation. This makes perfect sense in regard to the way humans interact with others and the environment, because it requires a certain amount of stability. Based on the crucial factors identified in earlier chapters, this chapter has defined innovation as a "change of beliefs that leads to a change of behaviour". It argues that innovation should be seen as a process of reconceptualisation rather than the generation of technological know-how or economic growth. This brings it closer to the understanding found in sociology, geography and archaeology and a little further from economics and political sciences. This bias has been justified through the disadvantages of the economic and political perspectives. This mirrors the sociological *Weltanschauung* upon which this research is based. Innovation as the recognition of new opportunities or threats can be stressful because it rejects strategies that have worked in the past, are socially accepted and possibly culturally embedded.

5. APPLICATION OF THE MODEL

Any useful model must represent the behaviour of real people in a real environment in a meaningful way. It has to be tested in real-life situations in order to judge its usefulness. This study developed a conceptual model of innovation with the aim of providing an unambiguous basis for interdisciplinary research and discussion. This constitutes a high goal and the remaining part of this book will try to prove that this aim has in fact been achieved. In order to do so the model will be applied to a case study which could – in theory - be set around any complex problem concerning social change. The model was not constructed in order to specifically approach archaeological problems, even though it will be applied to the archaeological record; if the model really describes the essential mechanisms underlying every innovation it can, in principal, be applied to any other innovation. This chapter will discuss ways in which the model can be tested within an archaeological setting.

5.1 PROBLEMS WITH TESTING THE MODEL IN THE ARCHAEOLOGICAL RECORD: Pots and Culture or how to link the present to the past

"Archaeologists want to make statements about behaviour, economic and social structures and so on which go beyond the data and are not themselves observable" (Hodder 1995: 83)

Recall that innovation is defined as an 'information flow that challenges beliefs and leads to a change of behaviour'. To test this model in the archaeological record requires knowledge about the beliefs (cause) and behaviours (effect) of people in the past. Neither is fossilized. While the latter is observable as changes in artefacts, archaeologists have very little direct evidence for changes in beliefs. Unlike other social sciences archaeology has no access to the beliefs of the people it studies through written texts. Though some of the later periods provide us with written evidence to supplement archaeological artefacts, these sources almost exclusively represent the thoughts of a highly educated élite and have probably been altered to glorify events. Therefore artefacts remain the main source of information of everyday processes.

Because belief systems are an essential part of cultural processes and influence many other aspects of human life, archaeologists cannot create a valid picture of the past without making assumptions about beliefs. They have to find ways to derive this knowledge from the artefacts they find. Dynamic cultural processes of the past have to be reconstructed indirectly from static artefacts and skeletal remains found in the present. This reconstruction process is complicated by differential preservation of artefacts which can alter their original composition. Archaeology therefore faces one basic problem: How should we interpret and analyse archaeological artefacts in order to gain valid information about the past? This question has triggered much debate over the last centuries (Johnson 1999: 14). The following section will briefly summarize the views on the 'correct' interpretation of artefacts that have been discussed in Britain from the middle of the last century.

5.2 POTS AS AN INDICATOR FOR CULTURE: Paradigm shifts in archaeological theory

During the 1920s and up to the 1960s archaeologists treated pottery and the people that used them as almost identical. Different pottery styles – like the Bell Beaker or the Linearbandkeramik – were the key variables to identify cultural groups (Hill 2002a). "We find certain types of remains – pots, implements, ornaments, burial rites, and house forms – constantly recurring together. Such a complex of associated traits we shall term a 'cultural group' or just a 'culture'. We assume that such a complex is the material expression of what today would be called a 'people'." (Childe 1929: v-vi). During that period archaeological research focused on the classification of artefacts according to their chronological order and spatial distribution. Based on this information cultures were defined with clear spatial and temporal boundaries. Changes within artefacts were generally explained with the invasion of foreign groups as the single cause of social change.

In the 1960s this *traditional culture paradigm* was challenged by the so-called *processual* or *new archaeology*. Its main accusation was a lack of understanding of cultural processes. Lewis R. Binford - one of the leading figures of the new approach - argued that archaeologists should aim to understand the archaeological record rather than purely describing it. The underlying assumption is that material culture is correlated with other aspects of social, technological and economic organization (Johnson 1999, Sabloff 2005: 212-219). This clearly widened the understanding of culture.

In the 1970s Binford introduced the so-called middle-range approach to archaeology. 'Middle range theory' was developed by the sociologist Robert K Merton (1968). Merton claimed that the nature of social phenomena does not allow for time invariant and all-embracing theories like in the natural sciences. He suggested that sociologists should not concentrate on empirical small-scale case studies, but should develop middle range theories that fill the gap between raw empiricism and all-embracing theory. Binford modified this concept, suggesting that the observation of contemporary cultures that live in conditions and ways similar to those in the past can provide a valuable technique to overcome the limited nature of the archaeological material. Such ethnological studies can provide us with analogies, which - given certain environmental and cultural circumstances – can explain what kind of behaviour has led to the traces observable in the archaeological record today (Binford 1983, Johnson 1999: 49-59). Experimental archaeology has also been used to reconstruct past behaviours. Both methods use the present as a model of the past and thus hope to extent our knowledge beyond the directly observable. For Binford the middle range approach is a link between the archaeological data and past culture processes or a means of interfering past behaviours from the archaeological record (Trigger 1997: 361-367).

Even though this has helped to provide valuable insight into past processes, middle range theory does not allow us to formally judge between alternative hypotheses. One difficulty when interpreting the archaeological record is that more than one behaviour might have generated the distribution of artefacts we observe in the present. Today it is generally accepted that the same artefacts (e.g. a particular type of pot) may have been used for different purposes depending on economic and social circumstances (Tyers 1996) and we therefore cannot gain certainty over which behaviours led to the artefact distribution we observe today. Hodder has defined this problem as *'equifinality'* (Johnson 1999, Renfrew 2005). This criticism, among others, led to the development of *post-processual* or *interpretive archaeology* in the late 1970s and early 1980s. It claimed material culture to be meaningfully constituted. This means that archaeologists have to understand the thoughts and values of the people they study if they want to reconstruct the past. Cultural ideas or beliefs influence human behaviour which consequently also affects the way in which artefacts get deposited in the archaeological record. In order to interpret artefacts 'correctly' archaeologists need to understand these beliefs (Johnson 1999). This last point brings us back to the original question of how to use the archaeological record to learn about changes of beliefs in the past.

While these theoretical discussions have widened the understanding of culture since the 1920s they could not solve the problem. Archaeologists are well aware of the insufficiencies of treating material culture as equivalent to the cultural processes they are embedded in. However the lack of more suitable scientific methods to even out the problematic nature of the material leaves them with few other options. A study of the role of pottery in social practice by Pitts (2004), for instance, compares the distribution of different form types from a range of sites in order to study intra-regional variation. The distribution of pottery is the observable effect of human behaviour but we do not know the exact causes of this pattern. By mapping the data Pitts tests whether there were behavioural differences between different regions. His research depends on the assumption that the behaviour (cause) that led to the particular distribution of artefacts (observable effects) was the same at all sites. But because humans can always choose between several behavioural options a certain variation in behavioural pattern can be expected, even within groups of the same cultural background. The results of such a case study can therefore be questioned on the basis of equifinality.

5.3 POTS AS AN INDICATOR FOR BEHAVIOURAL CHANGE

Analysis of variance could be used to detect innovation in the archaeological record using the conceptual model. Humans always have a behavioural repertoire that allows for a certain degree of reedoml; there is always more than one thing to do and different people might react differently to the same situation. This type of variation can be observed in the archaeological record as a variation of artefacts within one time phase. Since innovation is defined as a change of beliefs that leads to a change of behaviour, the replacement of one behavioural repertoire by another would suggest that innovation had taken place. In order to detect innovation in this way, variation resulting from the expected degree of freedom in human behaviour has to be clearly separated from the variation created by the introduction of a new set of behavioural norms. Analysis of variance works exactly this way; the statistical analysis can determine whether the variation between the time phases is too big to be explained by the variation within. This can be tested with the following two hypotheses:

> 0: The variation within phases is sufficient to explain variation between them. If this is true, there is no evidence of innovation.
>
> 1: The changes between phases are significantly greater than the variability within phases. This indicates a change of behaviour, which, if traced back to a change of beliefs, does represent innovation.

Various artefacts could be used as an indicator for behavioural variation, for instance coins or pottery types. If their variation is bigger between phases than within phases the null hypothesis is rejected and innovation took place.

However finding quantitative data for such a statistical test has proven to be difficult (McCarthy 1991). This might be surprising at first given that archaeologists generally accept the usefulness of quantified approaches (Orton & Tyers 1989). However, the nature of the archaeological record and the history of quantitative methods within the discipline can explain this lack of data (see more detailed discussion in chapter 8). Until the 1960s seriation was the principle quantitative method used by archaeologists (Tyers 1996: chapter 2). Then the paradigm change and the growing amount of quantified data resulting from scientific methods, like dating led to the use of statistical methods. In the beginning they were adopted from other disciplines, for instance cluster analysis used in biology or spatial analysis from geography. This led to criticism in the 1980s when it was pointed out that, because of the nature of the archaeological record, methods cannot be directly transferred from other disciplines (Orton 1992). Throughout the 1990s a number of projects investigated the possible application of quantitative statistics in archaeology (Winder 1996, 1997, Orton & Tyers 1989) but the methods have never found a wide application within the discipline. In general archaeologists do not seem interested enough to invest time to understand the arguments represented. While this might appear to reflect ignorance towards scientific methods it is rather the consequence of other methods being far more cost beneficial than statistics. Data needed to create distribution maps, for instance, requires far less preparation than data for quantitative analysis but it can still provide valuable information about variance in artefact distribution. Furthermore it cannot be expected that a statistical examination of data previously only examined in distribution maps or other formats would falsify existing results.

The use of unquantified data has demonstrated that archaeological artefacts can be used to find time-trends simply based on distribution maps and scatter plots. Hill (2002b) for instance has used vessel distribution and shape to demonstrate an increased variety in pottery between the Middle and Late Iron Age in Britain. His plots of rim diameter to vessel height show a significant increase in variation during the Late Iron Age, while the Middle Iron Age pottery shows a linear relationship with very little diversity. It is assumed that the Middle Iron Age pots had different functions but that is associated with surface finish rather than shape: burnished vessels were more often used for serving than cooking and plain ones more for cooking than serving. In comparison the vessel distribution from the Late Iron Age period shows a greater variability as well as a higher number of measurable vessels, which indicates that more complete vessels went into the archaeological record. These changes represent the effects of behavioural changes and the way potters thought about making pots. They went from using a single prototypical category to using a wider range.

Like other studies using artefact distribution maps and scatter plots, Hill (2002b) could demonstrate very clearly, that behaviour changed during the transition from the Middle to the Late Iron Age. However he and others could not provide evidence for changes of beliefs. These studies concentrate on information about the effects of human behaviour but do not provide us with information about its causes. They therefore give evidence of innovation in the sense of the economic understanding (technological change visible on an aggregated level of society, see chapter 2) but not according to the conceptual model suggested in this book. Traditionally archaeologists tend to explain one observable effect with another effect. Hill argues that the repertoire of tablewares increased during the Late British Iron Age because changes in table rituals raised the demand for new dishes. This leads to a chicken-and-egg argument. Does ceramic diversity cause table ritual or does table ritual cause ceramic diversity?

Existing knowledge of changes throughout the Late Iron Age and the Roman period (Cunliffe 2000, Millett 1990a: 30, Hill 2002b, Tyers 1996) provides sufficient evidence of changes in behaviour which can be considered to represent the effect of innovation. The challenge for testing the model will be to go beyond changes of behaviour manifest in the archaeological record and to find evidence for changes in beliefs. The applicability of the model depends on whether this can be achieved or not.

5.4 THE NATURE OF THE RESEARCH OBJECT

The discussion on how to observe changes of beliefs and behaviour in the archaeological record has revealed a cascade of difficulties. While changes of behaviour might be observed through changes in artefact distribution, the underlying intentions and cultural concepts remain inaccessible. This seems to be a general pattern in as much as the effects of innovation are easier to observe than its causes. As discussed in chapter 3 it is the knowledge about these intentions rather than the behaviour itself that will give us insight into human activities. In regard to the model, any observation of behavioural change is an insufficient indicator for innovation as long as it cannot be connected to a change of beliefs.

These problems are only partly due to the nature of the archaeological record and do not exclusively affect this discipline: They arise out of the nature of the research object. A change of beliefs can generally not be observed. This inaccessibility is a common problem in the social sciences, where many of the factors that are essential in regard to the research question cannot be observed. Cultural affinity, love, a virtuous act or sustainability are just a few examples. Even though they cannot be observed or measured it is generally accepted that they have an impact on human behaviour and play an essential part in the dynamics we study. Researchers can therefore not simply neglect their existence but have to find observable indicators or proxy measures. Like a conceptual model these indicators have to be justified and it has to be demonstrated that they actually represent what they are meant to.

A major part of the conceptual model of innovation suggested in this book cannot be directly observed which puts obvious limitations on its testability. The model will need to be expanded and the following chapter will discuss possible indicators that can be used to observe changes of beliefs indirectly.

5.5 CONCLUSIONS

In order to examine the usefulness of the conceptual model it will be tested in the archaeological record. The model describes the basic mechanism of every innovation and can thus provide a basis for interdisciplinary research. If this claim is correct, the model could be applied to any innovation and the choice of test scenario is a completely random one. Due to their limited access to the past archaeologists face a number of problems in reconstructing behaviours and thoughts. This chapter has discussed some of these problems. While certain difficulties, like biases through recovery techniques, partial preservation or equifinality, are caused by the nature of the archaeological record, others are not exclusive to the discipline. The nature of the research object, in this case human belief systems, does not allow a direct observation independent of the research setting. A testing of the model does therefore require a sensible indicator for changes in beliefs. The following chapter will discuss the options for expanding the conceptual model.

6. THE EXTENDED MODEL OF INNOVATION

This chapter aims to find an indicator for changes of beliefs that can be a) easily observed and b) theoretically connected to the conceptual model in a coherent way. This aim of connecting the archaeological record to human perception provides a classical example for a *middle range approach* in the understanding of Binford rather than Merton (see section 5.2). The next part of this book therefore follows the tradition of processualism.

So far the model has considered innovation in isolation, but humans do not act in a vacuum and are influenced by the environment which generally includes other actors and their intentions. Chapter 3 has shown how the environment provides incentives and constraints and thus influences human behaviour.

The model implies that every innovation starts on the micro level with one individual experiencing a challenge in beliefs that makes him recognise a new opportunity or threat. Information, even if passed on to a wider audience, is always perceived and evaluated by an individual and the same information can have very different meanings for different people. The starting point of every innovation is consequently located at the micro level while its effects can become visible on both the micro and the macro level. It is reasonable to assume that the majority of innovations remain on the micro level and go unnoticed by mankind. However every innovation has the potential to diffuse and thus cross the threshold and become a macro phenomenon. It is impossible to define the exact moment when this happens because it involves structural change. This diffusion process is also influenced by external factors. Whether the change of behaviour is accepted and adopted by the social group in which it is embedded depends on existing structures, social control and the receptiveness of people and institutions.

Both processes of the conceptual model - change of beliefs (cause) and the change of behaviour (effect) - act from the bottom up. The factors that constrain them are qualitatively different in that they act from the top down (see figure 7). Constraints generally influence the next level below in the hierarchical organisation of society, but Koestler has argued that higher levels can only interfere with processes on the lower levels to a certain degree. The conscious mind for example cannot be used to control functions on the cellular level (1967: 109). This perspective suggests two further aspects for an expanded model of innovation: the impact of top-down processes on its diffusion as well as the issue of scale.

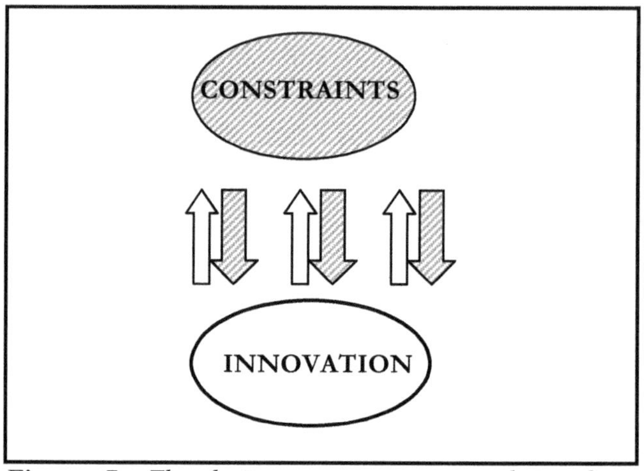

***Figure** 7: The bottom-up processes and top-down constraints of innovation*

6.1 TIME-SPACE CONSTRAINTS: Observing Innovation

"A changing, variable environment demands flexible behaviour, and reverses the trend towards mechanisation" (Koestler 1967: 111)

Even though the impact of other actors and the environment on innovative processes has been discussed before, it has not been integrated into the conceptual model. The next section will introduce the concept of *Time-Space constraints* (TS constraints) and explain why they can be used to account for the role of the environment (see box 2 & figure 8). This argument follows and widens the approach of the TiGrESS research group in Paris which investigates the development of city networks (Sanders et al. 2006). Cities are defined as individual agents and innovations are considered responsible for functional differences between them. Innovation is observed as changes in time-space constraints and measured as variation in the time needed to travel from one point to another. Like many other studies on innovation this example excludes innovations that have no impact on the macro level. A connection of TS constraints to the micro level causes of innovation will allow the observation of innovation on various scales.

The concept of Time-Space constraints can be linked to beliefs and behaviour in various ways:

1. TS constraints determine the option space of every actor. "Not only cognitive and social processes [...] but also solutions to technical problems are fenced in by the momentary availability of choice space" (Hägerstrand 1988: 220). *Changes* in these limiting factors consequently alter the degrees of freedom and create new conditions within which humans have to act. What an outsider observes as changes in TS constraints is experienced as a changing environment by the people affected. Such changes are likely to distract people from their routines, as old ways of doing things might no longer work. This might challenge the beliefs of an individual and even though not every belief challenge

necessarily results in a change of behaviour and thus innovation, a higher incident rate of its cause naturally increases the likelihood of innovation occurring.

2. Time and space play an important role in the transfer of people and information. As distance between places creates a barrier to human interaction, innovations that alter the way and speed at which people can travel also modify their space of social interaction. Hägerstrand (1988) has highlighted the role of geography in innovative processes. Using the example of a small island where two innovations spread slowly through personal communication until they eventually meet, he argues that once both are available in the same area they might be combined into something new. Therefore the speed with which information can travel has an impact on the probability space of innovation. During the last 2500 years the speed of transportation increased constantly. While an individual could travel an average of 5 km an hour within the city of London in 1800 the travel distance increased tenfold over the next 200 years. Before the innovation of electronic communication methods the speed of travel determined the speed at which information could be passed on. The possibilities and speed of physical mobility have increased significantly and today information flow is hardly limited by distance anymore. New opportunities in the distribution of information and more frequent encounters with new belief systems triggered by the interaction with people foreign to one's own culture are seen as factors that increase the likelihood of innovation.

3. Whether a micro level innovation diffuses and the change of behaviour is accepted and adopted by the group also depends on the receptiveness of people and institutions. Changes of behaviour on an aggregated level are very likely to have an impact on the whole society because they generally aim at change in the regulatory milieu. People can be expected to have a strong opinion on whether they favour such changes or not. Traditionally people in high positions have an interest in maintaining the status quo which put them into their powerful position and will only favour changes that will not threaten that position. These people respond to the challenges posed by potential innovations by increasing constraints on processes on a lower level and thus blocking innovation. This might cause a significant *time lag* between cause and effect (see box 2). When social conditions change, an innovation that was suppressed by old constraints may diffuse once these factors are eliminated. Change on the aggregated levels of society tends to be slow (see section 6.4) and the diffusion might take a significant amount of time. A similar argumentation can be found in the economic literature. In the discussion of economic long waves Perez (1983) has argued that the relationship between new technologies and the institutional framework is a crucial factor in regard to economic fluctuations. She claims that periods of great economic growth are characterised through a 'good match' between the requirements of a new technology and the institutional framework in which it takes place, while recessions are caused by a breakdown in the complementarity between these two factors. In other words, the social constraints within a society must allow an innovation to manifest itself in order for it to produce economic growth.

Changes in Time-Space constraints manifest themselves as changes in the environment. They create new conditions which are likely to stimulate innovation because they may challenge existing beliefs. These ideas are added to the conceptual model (see box 2). Because the moment an individual perceives a new threat or opportunity cannot be observed, changes in Time-Space constraints are used as an indicator. It is important to note that the relationship between TS constraints and beliefs is *not deterministic*; because innovations also occur in static environments and not every change in TS constraints leads to a change in beliefs. However TS constraints have an impact on the *probability space* of innovation within a *population*[1]. They affect the environment and alter the likelihood of innovation happening within a population. This argument leads to the following two hypotheses:

1: When the constraints acting on an individual are strong and the environment stable, normative behaviour can be expected, whereas

2: Once the constraints are relaxed innovative behaviour is more likely to occur.

Bottom-up or top-down? An ongoing debate

The expanded model can now account for both sets of processes that determine the human condition: bottom up processes that constitute or change social structures and top down processes that influence the choice of human action. While sociologists acknowledge the existence of both, theories traditionally tended to focus on only one side of this relationship (Dawe 1970, Archer 1982, Barrett & Fewster 2000). Dawe (1970) has described this separation of perspectives as the *two sociologies*: "One views action as the derivative of system, whilst the other views system as the derivative of action" (1970: 214). Early theories, like functionalism and structuralism, tend to overemphasise the role of social structure while interpretative sociology and hermeneutics neglect the role of structures (Archer 1982, Giddens 1984: chapter 1, Barrett & Fewster 2000).

Durkheim wanted to establish sociology as a science, and based on existing laws in the natural sciences, claimed that sociology needs to find similar laws for society. He formulated a mechanism of social systems like scientific laws. One of the regularities in the social world is the repetitiveness of human action, often based on compulsion. He concentrates his theory on the way in which society shapes the individual (Durkheim 1885: 105-140).

---

[1] Changes in the probability of innovation occurring always refer to a population or group rather than an individual.

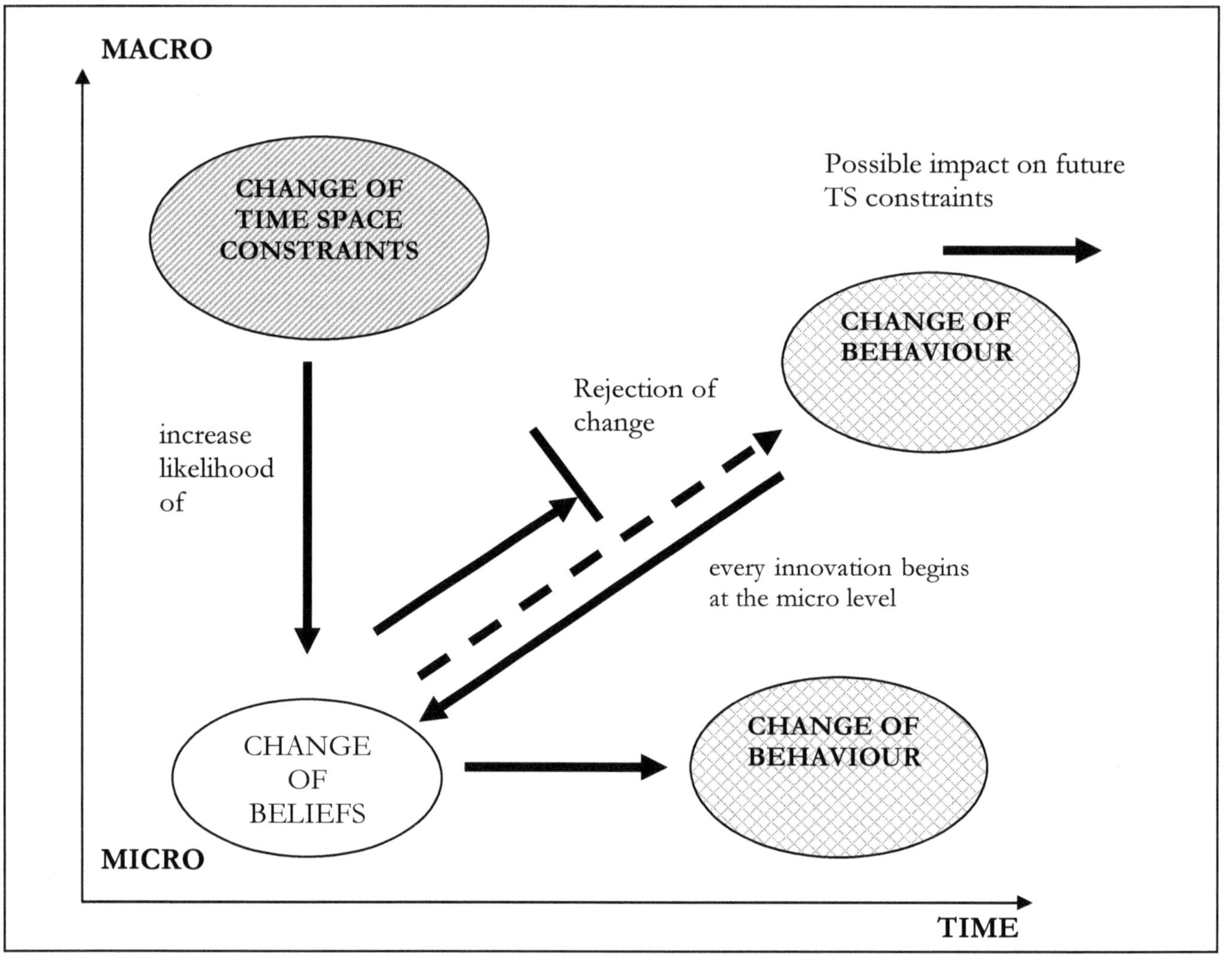

BOX 2: *The expanded model of innovation. Influences from the environment are represented through Time Space constraints, which can both increase the likelihood of innovation by initiating changes in beliefs and prevent innovation diffusion to a higher level of societal organisation.*

Marx' conclusion that "it is not the consciousness of men that determines their being, but, on the contrary, their social being that determines their consciousness" (1964: 51) sees human values and beliefs as determined by economic infrastructure. Bourdieu (1983) acknowledges that human action influences social structures (bottom up), but his main achievement is his work on habitus, the concept of incorporated and thus reproduced social structures (top down).

During the 60s there were attempts to reunite these two perspectives within one theory; the three logics by Esser (1993), the morphogenetic approach that originated in general system theory (Archer 1982) and Giddens' structuration theory (1984). The latter two were introduced in chapter 4. Esser (1993: 1-140) describes three different logics. 'The logic of selection' and 'the logic of the situation' describe how human action is influenced by structures.

The logic of the situation is influenced by the type of logic of selection chosen by the individual, for instance whether it triggers an emotional outburst or a rational consideration of various alternatives. The resulting action collectively creates what Esser calls the 'logic of aggregation'. This step describes the other half of the cycle, how action can create, alter or preserve social structures.

The principles of 'top down' and 'bottom up' processes have also been described by various other disciplines as the basic mechanisms of change and stability: In evolutionary systems variation is spontaneously created on the micro level and filtered by natural selection which can endorse or reject the emerging pattern (Darwin 1959). Hägerstrand's Time Geography studies agents moving through time and space as they try to realise a project under the restriction of Time-Space constraints.

Even though all these disciplines talk about the same mechanisms, many of them only concentrate on one aspect. The different labels used to describe these processes also conceal this fundamental similarity. This is a remarkable aspect of innovation: although definitions falsely fuse very different viewpoints and processes, its underlying hierarchical mechanism is widely accepted.

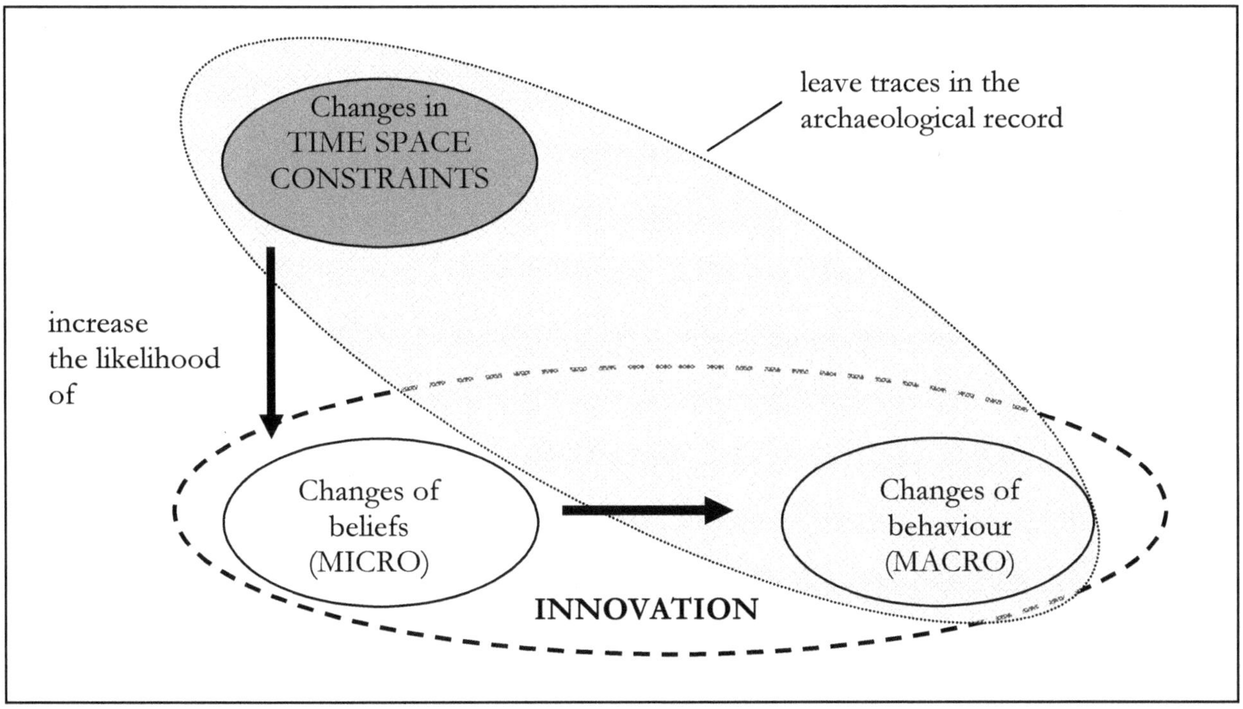

Figure 8: *Schematic representation of the methodological framework: The concept of Time-Space constraints allows an indirect observation of changes in beliefs*

6.2 TIME-SPACE CONSTRAINTS: Cause, effect and multiplier effects

While changes in TS constraints increase the likelihood of innovation, innovations can also modify time-space constraints, even if they are not intending to do so. The relationship between the two is interlinked in as much as changes in TS constraints caused by an earlier innovation may increase the probability of future innovations taking place (box 2). These processes are never ahistorical and the relationship between innovation and changes in TS constraints is not circular[2]. In some cases a technological innovation has to precede others to make them possible. Some changes in TS constraints predate a particular innovation, which transforms them and these new constraints logically postdate the innovation. While the development of the telephone did not cause the invention of skyscrapers, it provided a new means of communication that removed limitations and thus made skyscrapers possible (Hägerstrand 1988).

Innovation can underlie a 'multiplier effect', a term used by Colin Renfrew in his book *The Emergence of Civilization* (1972: chapter 3). Multiplier effects arise as a result of 'positive feedback loops' or reinforcing mechanisms. Feedback loops can be positive or negative, a distinction that belongs to cybernetic theory.

While negative feedback loops tend to buffer or suppress change, positive feedbacks amplify them. They tend to move a system away from its equilibrium and make it more unstable. In a negative feedback loop, change in one variable (increase or decrease) leads to the opposite change in a second variable. In a positive feedback loop the system responds to the change in one variable with the same type of change (increase or decrease) in another variable. An example would be the world population with a fixed birth rate: the population will grow proportionally as a large population leads to large numbers of births and large numbers of births result in a larger population. If such positive feedbacks are not counterbalanced by negative feedbacks they can run out of control.

We can easily see how such loops arise for innovation (see box 2). A response to a threat or opportunity leads to a change of behaviour, which may create new threats or opportunities that stimulate further innovations, leading to a *cascade of innovations* on many spatio-temporal scales. TS constraints increase the likelihood of innovation occurring and at the same time often occur as the result of innovation.

The model now covers two levels of social organisation (see box 2). Because the moment that an individual perceives a new threat or opportunity cannot be observed, changes in Time-Space constraints are used as an indicator. They increase the option space of individuals. TS constraints are observable within the archaeological record and can be linked to distribution maps of artefacts, which represent changes in behaviour. Furthermore the expanded model allows for observation of large-scale innovation, whilst it additionally provides an explanation by including the causal mechanism on the micro level. Disciplines that focus on macro changes do not have to consider micro processes in order to observe innovation, but can use the micro level for additional information.

[2] If X causes Y, Y cannot simultaneously cause X. The relationship between Time-Space constraints and innovation develops over time.

6.3 STRUCTURES AND CONSTRAINTS

The concept of Time-Space constraints is very similar to the role of social structures discussed in chapter 3: both influence human actions and can be altered through them. The following section will discuss a) why, despite obvious similarities, two separated terms are used to describe these processes and b) why the concept of constraints is used to expand the model rather than social structures.

Both structures and constraints cannot be observed directly. They only become visible through the actions of individuals influenced by them. While a watering-hole, for instance, still exists when nobody uses it, a social norm only materialises through its influences on human behaviour – that includes conscious disregard by the actor. While we can read a law or observe a new road being built, their impact on society only becomes visible through the way human behaviour is influenced by them. It is crucial to point out that social structures are culturally constructed and highly dependent on time and space, even though the individual itself might experience them as an objective universal truth which exists outside human agreement (Schimanck 2000: 15, Bourdieu 1988, Garfinkel 1973). "Structural properties of societies are real, but at the same time they have no physical existence. They depend on regularities of social reproduction" (Giddens & Pierson 1998: 77).

There is an important qualitative difference between structures and constraints: while by definition any kind of constraint reduces the option space of an individual, structures can be enabling as well as restricting (see figure 9). Some sociologists regard structures as a constraining force over human action. Durkheim (1885) has emphasised the way in which structures mould human behaviour into a socially accepted form during socialisation and thus adapt an individual to society. Others have considered structures to be constraining as well as enabling, based on the argument that every choice of action excludes other alternatives but at the same time offers certain opportunities (Archer 1982, Giddens 1984: 117, 169 - 174, Schimanck 2000: 19). Giddens explains this through the example of learning a first language; this process is determined by structures and while it limits our ability for other languages it widens our cognitive and practical capacity and thus enables us to become a member of our society (1984: 117). Giddens criticises Durkheim's perspective as well as Hägerstrand's Time Geography, for solely focusing on the constraining aspect of social structures (1984). He stresses that "all types of constraints [...] are also types of opportunity, media for the enablement of action"[3] (Giddens 1984: 117). Of course the matter of constraint or opportunity is highly dependent on perspective; one person's chance can be another person's constraint. A law that claims that every child between the age of 6 and 15 must go to school can be seen as a constraint: the child is not allowed to work full time; or as an opportunity to learn. In the same way individual behaviour constitutes social structures and constraints, it is human perception that determines whether they are considered to be a constraint or an opportunity.

Structures are not the only influences on human action, but sociologists tend to ignore other factors (Hägerstrand 1975, Giddens 1984: chapter 3, Schimanck 2000: 14). Physical factors, the physical state of the individual as well as the weather or other environmental factors can constrain or enable the space of potential actions, but sociologists only account for them if the research interest makes it necessary. For instance, our death as a time-limiting factor is crucial to rational choice argumentations. If a human life-span was infinite we would all be able to do everything we want. But in real-life many 'goals' come with deadlines; the ability to reproduce, for instance, is limited by a woman's age.

Time-Geography acknowledges the constraints on human action that result from the nature of the human body and the physical context in which the action occurs. Hägerstrand (1975) identifies two different types of constraints, the first one groups circumstances which cannot be improved by scientific or political interventions and the second type contains constraints which are, at least in principle, amenable. Hägerstrand names a number of constraints he considers to have a crucial impact on the freedom of human behaviour (1975: 12):

(1) Indivisibility of the human body
(2) The limited length of a human life-span which makes time a scarce resource and thus limits the adaptive potential of an individual
(3) The limited capability of human beings to participate in more than one task at once
(4) The fact that humans cannot move in space without moving in time
(5) The fact that every task consumes time
(6) The limited packing capacity of time-space, no two individuals can be in the same space at once
(7) The history of events, every situation is rooted in past situations

Although this list seems to focus solely on 'time' and 'space' the constraints suggested by Hägerstrand also include cultural aspects. While humans cannot move in space without time, technology and innovations have decreased the amount of time that is needed to move around. Movement can also be restricted by social laws rather than geographic factors. Today the crossing of borders between nations is restricted by nationality rather than the characteristics of the landscape. Thus the concept of TS constraints can account for social constraints, while social structures ignore physical restrictions. The latter

[3] While Giddens' choice of words is certainly unfortunate and imprecise - the term 'constraints' simply implies constraining and not enabling - his argument summarises the sociological understanding of the impact of structures on individuals.

term covers a more detailed list of factors that have an impact on human activities.

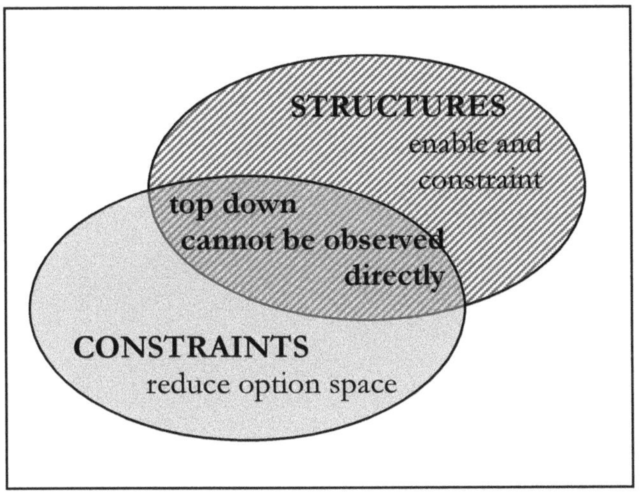

Figure 9: *Differences and similarities between the concepts of constraints and structures. Both operate at an aggregated level, but while structures can be enabling as well as constraining, constraints solely reduce the adaptive potential of an individual.*

Chapter 2 has demonstrated how the complexity of social systems prevents predictions about the direction they will take. Schimanck (2000: chapter 7) claims that social dynamics or structures can only be theoretically reconstructed when they are closed rather than open. Consequently sociologists cannot describe complex social problems. Since the vast majority of social dynamics are open or a mixture of open and closed processes, sociologists cannot make any sensible predictions about the future of social structures. While social structures are a useful theoretical concept, they do not provide the basis for a method. This supports the claim that in regard to capturing human activities the concept of constraints holds certain advantages over structures. Hägerstrand claims that the researcher should "find out in what ways limits of freedom of action come about and who, if anyone, is responsible for the configurations of such limits" (1975: 3). The focus on constraints has the advantage that it allows the prediction of the direction of human action. If the constraint is strong the option space of an actor is very limited and the spectrum of possible action is predictable. This situation is comparable with the normative action model; if we know the norms and know that an individual will follow them, we can predict what it will do.

It has been argued that in periods where constraints are strong unpredictable (innovative) behaviour is less likely to occur. Archer (1982) has argued that the potential for change is rooted in systemic stability or instability. She distinguishes between 'hyperactivity', where all rules and resources are defined as transformative and actors enjoy a very high degree of freedom, and periods of 'rigid structures', when structural properties are reconstituted through human action.

Even though the conceptual model developed in this study was based on the impact of social structures on the human condition, it will use the concept of Time Space constraints to represent the impact of the environment. This has three reasons: firstly the term 'structure' tends to ignore the impact of physical space and time while constraints also include socially constructed constraints. Secondly, by allowing for factors that enable human action *as well as* factors that constrain it, structures might appear to cover more of the processes that define the human condition. However the definition of what is an opportunity and what is a constraint lies in the eye of the beholder. Thirdly constraints allow a less unambiguous bounding of the research question. A constraint always limits the space of potential actions and thus makes human behaviour more predictable. However it has to be noted that *changes* in TS constraints often provide new opportunities and the suggested focus on changes consequently includes the aspect of the environment as enabling as well as constraining.

6.4 SCALE AND PREDICTABILITY

"Wherever there is life, it must be hierarchically organised" (Koestler 1967: 47)

The literature review (chapter 2) has shown the majority of innovation research focuses on changes observable on the macro level. Most economists and policy makers are only interested in innovations that diffuse and become visible on a higher level of social organisation and thus have an impact on the course of history. Archaeologists are interested in individual processes but are unable to observe small-scale changes due to the nature of the archaeological record. One of the reasons for expanding the original model is to provide means to observe macro innovations. The diffusion process takes place over various temporal and spatial scales which has implications for the predictability and management of innovation. Processes on different scales do not have to be harmonized and in fact often conflict each other. It is, in fact, this link between the micro and the macro level that constitutes the central problem when dealing with human societies and complexity. Individual behaviour might result in emergent group-level patterns, whose properties are impossible to predict and irreducible (Holland 1998, Koestler 1967).

The aspect of scale is connected to *rates* of change, which vary between different levels of social organisation (Ogburn 1966, Ellen 1994). There are different views concerning at which level of a society change occurs fastest. While Ellen argues that "we can hypothesize change taking place more slowly at the level of domestic reproductive behaviour compared with the level of regional systems" (1994: 70) other authors claim the opposite by saying that "roughly speaking adjudication at the level of behaviours has a time scale of years [...] and adjudication at the level of cultural constructs at the level of hundreds of years" (Read 2003: 38). Generally processes on the lower levels of observation are completed in shorter time periods than processes on a higher level (Ahl & Allen 1996). Differences in the rate of change are crucial for any framework which claims to examine social and cultural change in as much as it must be able to account for them. Changes in one area of a social system might result in unexpected consequences in another and the different rate at which these changes occur might lead to conflict. A new technology, for instance, might spread before the social system can adapt to the changes it entails. It has been suggested that technological changes tend to proceed more rapidly than changes in other aspects of culture (Ogburn 1966, Ellen 1994). "Technology and production, on the whole, change more quickly than political relations and ideologies, which carry with them cultural baggage from the past which is not easily disposed of" (Ellen 1994: 67). The rate of change is also crucial in regard to observation, which is limited by the factor time. A short observation time does not allow the observer to fully monitor certain processes. By watching an individual for a day it is not possible to say if his or her behaviour is habitual or unusual.

The degree of behavioural freedom also depends on scale as we have "[...] the appearance of more complex, more flexible and less predictable forms of behaviour on successively higher levels of a hierarchy" (Koestler 1967: 104). According to Koestler improvisations or non-fixed actions are only possible on higher levels. It is possible for behaviour to move down levels. Often when we first practise a new kind of behaviour we have to concentrate on every move but once it turns into habitus and we can act automatically.

The same factor can have different results depending on the scale it is influencing. Stress has been discussed as a possible cause of innovative behaviour (see chapter 2). However this assumption is not generally true but scale dependent; while stress experienced by an individual might stimulate a new way of seeing the world, global stress can actually prevent institutions from capitalising on new knowledge and thus prevent change of behaviour. The aspect of observation scale is also crucial in regard to the previous discussion on structures and constraints. Only the individual affected by certain structures can perceive their enabling nature. An observer can only observe the same structure as an constraint on the macro level.

So far innovation research has shown very little effort to relate the processes observed at different scales or even to acknowledge the involvement of different processes and mechanisms. Innovation is very much seen as one process. Hierarchy Theory (Ahl & Allen 1996) is an approach that considers the consequences arising from different levels of observation and its impact on the understanding of the research object. It is a branch of general system theory and has been defined as "a theory of the observer's role in any formal study of complex systems" (1996: 29). It acknowledges that the choice of question has an impact on the answer as well as the mode of explanation. Hierarchy Theory claims that "in order to describe a complex system adequately several levels need to be addressed simultaneously" (Ahl & Allen 1996: 30). It claims that predictability is only possible in complex systems when many scales are taken into account simultaneously.

6.5 CONCLUSIONS

The conceptual model defines a change of beliefs as the cause of innovation. Because this process cannot be observed directly, this chapter has suggested *changes in Time-Space constraints* as a possible indicator. Time-Space constraints represent the impact of the environment on the option space of each individual. Changes in these constraints are experienced as changes in the environment and behavioural routines might become invalid within these new circumstances. The likelihood for an individual to perceive a new opportunity or threat is higher in such a situation than in a static environment. Thus changes in TS constraints *increase the likelihood* for innovation to occur, even though the connection is by no means a deterministic one. This argument leads to two hypotheses: 1) When constraints on the individual are strong and remain stable normative behaviour is expected, whereas 2) when the constraints are weak and the environment variable, innovative behaviour is more likely.

This chapter followed a middle range approach as suggested by Binford. The introduction of TS constraints allows a connection between the archaeological record and human perception. However the concept of Time Space constraints is in principal also applicable to other social science problem domains that include innovation. This work has also linked the sociological understanding of the human condition to the concepts of Time Geography in order to gain an understanding of the archaeological record.

The expanded model contains the two processes that constitute the human condition: top-down and bottom-up. It also allows the observation of macro scale innovation, whilst it additionally provides an explanation by including the causal mechanism on the micro level. Disciplines that focus on macro changes do not have to consider micro processes in order to observe innovation, but can use the micro level for additional information.

This chapter also introduced the possibility of innovation cascades, in which one innovation creates fertile conditions for further changes until the old conditions have almost vanished. It also discussed the impact of scale on innovative processes and has demonstrated that innovation cascades often occur on a range of spatio-temporal scales.

PART 2 – CASE STUDY

Theory evaluating

INTRODUCTION

The first section of this book developed a conceptual model of innovation with the aim of providing an unambiguous basis for interdisciplinary research and discussion. This constitutes a high goal and the remainder of this study will try to prove the usefulness of the model by applying it to the Romanization of the North-Western Provinces from about 50 BC to AD 50. While the choice of discipline was a matter of convenience, there are a number of large-scale innovations manifest in the archaeological record which could have been used to test this model, for instance the spread of the Neolithic revolution or the first usage of bronze. All constitute changes in behaviour that had major impacts on the living conditions and long-term effects. The test scenario was chosen in regard to hypothesis generation. The region and time period studied here cover areas that were already conquered by the Romans (Gaul), areas which were not yet under Roman control (South East England) as well as areas which would never be (Germany north of the Rhine, Scotland). This allows a comparison of areas before and after Roman occupation, as well as between conditions in which innovation took place as opposed to conditions under which it was rejected. Such a comparison is likely to provide additional insight into innovative processes.

The limitations of the archaeological record in giving evidence of human behaviour and its underlying intentions have already been discussed. It could be argued that if new insights can be gained under such imperfect conditions, the model should be even easier to apply to more contemporary questions and in disciplines that have more direct access to human behaviours. What is important is that the model was built in one set of data (literature on innovation) and will be tested in another, completely independent set of data (the archaeological record). This is an unusual approach for archaeology where models are often built based on patterns observed in the archaeological record and are then tested in the same scenario. This always bears the danger that we "observe what we believe and then believe in that which we have observed" (Clarke 1972: 6). Due to the way humans make sense of their environment, scientific models are always influenced by the cultural background and knowledge of the scientists. The history of paradigm shift (Kuhn 1996) shows that major changes in the scientific understanding of the world were always accompanied by social change. This cannot be avoided; it is part of the human condition. One way to minimize the impact is to make the model used explicit. This study strictly divides theory building from theory evaluating in order to gain some independence between theory and evidence. At the same time this second part will try to connect the model with some data to demonstrate its usefulness.

Connecting the model to an archaeological case study

The application of the model to the Romanization of the North-Western Provinces requires evidence for changes in Time-Space constraints (indicating changes in beliefs) as well as evidence for changes in behaviour. There are a variety of archaeological artefacts that could be used to observe changes in behaviour; chapter 5 has demonstrated how patterns in pottery distribution and shape have been used for this purpose. This case study will be based – again rather randomly – on the analyses of the faunal remains of the main domesticates, cow, sheep and pig, for the British Iron Age. Studies about subsistence strategies of past societies do not only provide us with information about the food consumed but have far wider social, cultural and environmental implications. Faunal evidence provides us with information about the ways in which food is linked to social and economic factors, while its preparation is linked to gender relations, labour division and cultural beliefs (Meadows 1994, Cool 2006).

The faunal remains will be examined for evidence of *changes in behaviour*, which has been defined as the observable effect of innovation. Results of previous studies using faunal remains to understand the Romanization of Britain will be considered and reassessed. In a second step the pattern will be connected to changes in Time-Space constraints. Remember that the first part of this book suggested a close link between innovation and TS constraints: when the constraints on the individual are strong and remain stable normative behaviour is expected, whereas when the constraints are weak and individual behaviour meets only little restriction, innovative behaviour is more likely. Evidence for changes in TS constraints therefore indicates that *changes in beliefs*, which provide the cause of innovation, were likely. The introductory literature review on the North-Western Provinces will focus on archaeological evidence for such changes.

It is generally accepted that the process of Romanization provided new opportunities and threats for the insipient populations. Roman creations like the administrative system, laws, a citizenship for an area as vast and heterogeneous as the Roman Empire or new infrastructures are without doubt significant alterations. There are two gravestones from the 2nd or 3rd century AD in South Shields, North East England that show Syrian elements and give evidence of the long distances that at least some individuals travelled during Roman times (Phillips 1977: gravestones 247 & 248). New roads built by the Roman army as well as the abundant usage of waterways had effects on such various aspects of life as trade, military action and communication (Chevallier 1976, Peacock & Williams 1986). The military roads could be used by traders as well and the Roman Empire saw people moving around like never before in history. Transport opportunities provide the physical limit of trade and its cost can have an impact on the structure of agriculture (Greene 1986: 43). In antiquity the speed by which people could travel resembled the speed by which news could be transmitted (Chevallier 1976: 191). These few examples show that even though Time Space constraints describe geographical factors, they influence our lives in various social ways. What an outsider observes as changes in TS constraints is experienced as a changing environment by the people concerned.

7. DEVELOPMENTS IN THE NORTH WESTERN PROVINCES OF THE ROMAN EMPIRE FROM 50 BC TO AD 50

The process of Romanization has experienced a lot of attention from archaeologists and historians and the transition from the Iron Age to the Roman period is considered to be a period of major change.

The next section will offer a brief summary of the archaeological debate on Romanization and how this process should be examined.

7.1 THE CONCEPT OF ROMANIZATION

The concept of 'Romanization' was first developed by Francis Haverfield and Theodor Mommsen at the beginning of the 20th century (Webster 2001, Mattingly 2004). Initially Romanization was studied from an imperialistic point of view. The changes it brought about and that resulted in the "native culture more closely resembling that of Rome" (Millett 1990a:1) were explained as a one sided acculturation of the new provinces brought about by Roman presence. Romanization was characterized exclusively through top-down processes (Meadows 1994, Webster 2001, Hingley 2003).

Later this perspective was criticised for its Roman bias and for rejecting the role of the indigenous populations during the transition. It has been argued that Romanization should be seen as dialectic change (Webster 2001, Millett 1990a & b, Meadows 1994). The conceptual model supports this understanding of social change and regards innovation as a multifaceted process with causes and effects on various levels of society. The understanding of the human condition suggested in chapter 3 makes a one-sided acculturation implausible, because most changes contain bottom-up as well as top-down processes.

In recent years the debate about Romanization has shown a strong emphasis on the aspect of 'identity' and what it meant to become Romanized (Webster 2001, Hingley 2003, Mattingly 2004). This perspective disagrees with the assumption that "everyone in Roman Britain would have recognized the innate superiority of Roman civilization and have wished to subscribe to its material comforts to the maximum extent possible" (Mattingly 2004:6) and argues that we should gain a better understanding of what Romanization meant to the majority of people and their everyday life, including aspects of resistance and the merging of cultural concepts.

However the first part of this book has discussed the difficulties of assessing people's beliefs systems, a disadvantage that poses greater problems to archaeology than other disciplines due to its dependency on static artefacts. This case study does not therefore aim to contribute to the debate on 'identity' or 'ethnicity' in regard to Romanization. It has introduced the concept of Time-Space Constraints as an indicator for changes in beliefs due to the inaccessibility of this aspect of the human condition. While it is hoped that the application of the conceptual model will provide new insight into the process of Romanization of the North-Western Provinces, the set-up will not provide hypotheses about the exact nature of changes in beliefs systems that have occurred during this transition period.

Literature sources as well as a number of different artefacts have been used to study the transition from the Iron Age to the Roman period in Britain, for instance pottery (Greene 1978, Hill 2002b), settlements (Hill 1995, Cunliffe 2000) or faunal remains (King 1985, 1999, Hambleton 1999). Other approaches have also used parallels with other periods of major change which were characterised by one culture spreading to new territories, for instance the creation of African-American societies in the new world (Webster 2001) or globalization (Hingley 2003).

The rest of the chapter will give a brief overview of the events taking place in Gaul, Britain and Germany from 50 BC to AD 50 and concentrate on changes in Time-Space constraints that can be assumed to have significantly modified the environment of Iron Age people.

7.2 GAUL

Roman presence in Gaul began with a plea for help by the Greek city-state Massalia (Marseilles), which was threatened by the Celto-Ligurian tribe of the Salluvii (see figure 10). In the 120s BC the Romans stationed a legion at Aqua Sextia (Aix-en-Provence) and in 122 they successfully defeated the Allobroges who had supported the Salluvii. In 118 BC the first Roman colony was founded at Narbo Martius (Narbonne). It was linked via an important land road to Italy and Roman possessions in Spain. Despite the Roman presence Gaul stayed troubled: it was continuously threatened by Germanic tribes in the North as well as by revolts of Gallic tribes, the most serious of which was the rebellion among the Allobroges which was quelled in 62-61 BC.

The 60s BC saw a power struggle between Caesar, Pompey and Crassus in Rome. When the Helvetii, who were threatened by the Germans, attempted to migrate west into Gaul, Caesar took the opportunity to win military glory by campaigning in Gaul. Between 58 and 51 BC he conquered the territory of what is today France, Belgium and parts of the Netherlands, as far as the Rhine, and Gaul came under Roman domination. Caesar left an account of these wars in his *De Bello Gallico*. These seven years of war had an impact on the Gallic culture and left Gaul economically and socially damaged. Many Gauls were killed or enslaved during the wars and Gaulish gold disappeared into Roman property. However the defeat of the Helvetii gave Caesar huge prestige in Gaul (Cunliffe 2000: 434-442, Drinkwater 1983: chapter 1, Wightman 1985: 34).

In 51 BC Caesar left for Italy and never returned to Gaul. But even though Rome was distracted elsewhere and showed little interest in influencing political processes in Gaul the army remained in the provinces until it was moved north for Drusus' Rhine offensive in 12 BC. The reason for this continued occupation was that the Gallic tribes remained troublesome and periods of military withdrawal tended to be followed by unrest. The Germans kept on threatening the new Provinces and supported Gallic groups in their rebellion against the Romans. In 39 BC Agrippa became governor of Gaul and campaigned successfully to settle the unrest. However he had to return 20 years later because of new rebellion.

Before the construction of a Roman Road system in Gaul, the Romans used roads created by Iron Age traders. They had proven very helpful for Caesar during his campaigns (Chevallier 1976: 15). These pre-Roman roads were mainly geographical routes and followed the valleys of great rivers. The rivers in Gaul are distributed in such a way that they provide a great means of transport, as becomes clear from Strabo's comment "the course of the rivers is so happily disposed in relation to each other that you may travel from one sea to the other [...] ascending some, and descending others [rivers]" (West quoted after Chevallier 1976: 160). The Celtic Gauls were also highly-skilled vehicle builders and provided expertise the Romans could exploit (Chevallier 1976: 178, Greene 1986: 37). The exact date of the construction of the Roman road system is unclear, but it is generally attributed to Augustus and Agrippa (governor of Gaul in 39 and 20 BC). The road system was completed by Claudius (41-AD 54) (Drinkwater 1983: 38, Wightman 1977). The roads were constructed because the Romans needed supplies for their troops as well as the opportunity to move them quickly around the country to areas of conflict. Agrippa's military successes support the idea that roads were essential for the Roman troops. They are considered to have constituted the basis for a permanent control of Gaul (Simon 1982a: 38) and increased the speed of travel two- or threefold (Greene 1986: 35). Because of the short lifespan of food and the high cost of land transport it is likely that most of the supplies were provided from areas in close proximity to the army bases. Therefore the presence of Roman troops put stress on the native population, as it represented a sudden increase in people dependent on the labour of others (van Groenman-Waateringe 1980). The Roman Army also had an influence on native tribes because it used to recruit new soldiers from the local area and therewith withdrew young, strong men (Drinkwater 1983: 120).

While Roman military roads were of good quality and built to move legions and messengers quickly, the high cost of land travel made them less interesting for trade. Middleton (1979) has argued that the new roads will have had only a little impact on trade because of the existing and very effective river system in Gaul. Comparisons with 18th century England have suggested that the use of river routes was around five times cheaper than road transport (Chevallier 1976: 160-162, Drinkwater 1983: 127). From an economical perspective goods were more likely to be taken by water (sea and rivers) than via land (Peacock 1978). However, as Greene (1986: 39-42) points out, the number of trading points covered during a journey might also have been an important factor in the choice of trading routes in Roman times. Although the shorter route via the Mosel and Rhine would have been more expensive than taking the cargo via Gibraltar, the river route provided contact with major cities and military fortresses and could have been chosen for this reason despite its economic disadvantages. The road network (see figure 10) consisted of four major military ways, which centred on Lyon (Lugdunum) and from there followed the rivers Rhone, Saone, Moselle and Rhine. There is a clear connection between these roads and the waterways.

The Roman Army has also been held responsible for the development of long distance trade. The establishment of a monetized economy and the presence of soldiers who were paid in coins stimulated trade which showed a strong emphasis on military markets (Middleton 1979: 85). Around AD 21 there was a general money shortage and people in Gaul were complaining about the high taxation and the brutality of the governors. This led to a revolt by a few Romanized aristocrats and even though their rebellion was quelled by the Romans, these events revived the 'Terror Gallicus' and the impression that Gauls could not be trusted (Drinkwater 1983: 28f).

The Roman army must also have had a significant influence on Gallic potters. The so-called terra sigillata pottery, which originated in Italy, was later also produced in the provinces. These products were produced in order to satisfy the demands of the army and the elite. During the early Augustan period kilns were set up in Lyon and this development eventually let to the erosion of Italian wares and their replacement with provincial products. The new production centres had the advantage of access to the river network in Gaul. The amphorae found in Gaul were generally used to import wine from Italy and olive oil and fish sauce from Spain. Between 30 BC and 15 BC the archaeological record changes considerably: Roman coins become more abundant while native ones decrease and southern influences and imports become more visible in Northern Gaul (Greene 1986: chapter 3, Wightman 1985: 48 & 142).

The open border to Germany remained one of the main threats to Gaul. In 16 BC invading Germans killed part of the Roman Army in the so-called "Lollian massacre". Augustus visited Gaul in the same year and appointed Drusus as overlord for a great offensive against Germany which started in 12 BC. This campaign influenced Gaul as troops stationed in the centre of the country were moved to Lower and Upper Germany, leaving the hinterland relatively unsupervised. In order to avoid an uprising in the rear and to bind the Gauls closely to the Roman cause, Drusus built a great Altar just outside Lugdunum in the same year. Sacrifices at this altar honoured both Rome and Augustus. Meetings were held every year at this altar and sacerdos were elected. Their responsibilities are disputed but it has been assumed that they were

Figure 10: *Major events and cultural groups in the North-western provinces from 50 BC to 50 AD. Based on Drinkwater (1983), Simon (1982a & b), Cunliffe (2000) and Jones & Mattingly (1990).*

connected with the management of funds. These meetings constituted the *Concilium Galliarum*, the 'Council of the Gauls' (Drinkwater 1983: 111-122, Wightman 1985: 51). After the German conquest failed and the Roman army was pushed back from the Elbe to the Rhine (AD 9), permanent legions were positioned along the west bank of the Rhine, which had effects on the economy and administration of Belgica Gallia (Hamilton 1996).

7.3 GERMANY

The development of both Gaul and Germany were closely interlinked during the 100 years considered in this research. The area around the Rhine was conquered together with Gaul in 55 BC. Until 12 BC and then again after AD 9 the river constituted the border of the Roman Empire (Simon 1982a: 38) (see figure 10). Tribes in the remaining free part of Germany constantly threatened the peace in Gaul. In 16 BC the Germanic tribes Usipiter and Sugambrern defeated a Roman legion which initiated a Roman attempt to order the situation at the northern border. The initial plan was to create security for Gaul with German territory as a buffer zone in front of the border. In the same year the Romans erected the first known camp in Neuß and just before the attacks were started camps were built in Mainz (Mogontiacum), Birten

(Vetera) and Nijmegen (Numaga). They were occupied by soldiers that had served in central Gaul before. In 12 BC Augustus's son Drusus invaded Germany, starting off from these newly-erected bases. His first attacks were aimed at the tribes that had defeated the Roman legion 4 years before (Simon 1982a: 40-41, Drinkwater 1983: 22-25). In AD 1 most of the Germanic tribes between the Rhine and the Elbe revolted and it took the Roman troops a long time to restore the order created in the campaigns of 12 to 7 BC. These events led to a change in Roman policies and instead of using the newly-gained territory as a buffer the Romans started preparations for a Germanic province (Simon 1982: 49).

In comparison to Gaul, Germany was economically and socially less developed. The lack of infrastructure did not allow the Romans to winter their army above the Rhine, because they could not supply their troops with food (van Groenman-Waateringe 1980). Despite this problem Drusus reached the Elbe in AD 9 but did not cross the river. On his way back Drusus had an accident from which he died shortly afterwards and his brother Tiberius took over the troops. In AD 9 a revolt by Arminius destroyed three legions led by Varus at the Kalkriese battlefield. Considering that the entire Empire only had 28 legions at the time, this was a serious blow to Rome and the defeat resulted in a withdrawal of the Roman troops behind the Rhine and into the position of 12 BC (Simon 1982a: 42, Drinkwater 1983: 22-25).

Tiberius (14-37) became Augustus' successor and his father advised him to keep the Roman Empire within its existing borders. In the later years of his power, politics on the Rhine were concentrated on a stronger defence. Tiberius died in AD 37 and was succeeded by Gaius who was assassinated in 41. Sources mention a campaign against the Germans by him but the text is not very clear. It is stated that Gaius suddenly decided to campaign in Germany and had crossed the Rhine without moving far into Germany. A conspiracy against the emperor is mentioned as a possible cause. Gaius died shortly after returning from Germany and was replaced by Claudius (41-AD 54). His politics regarding Germany were overshadowed by his invasion of Britain in AD 43 (Drinkwater 1983: 54-56, Simon 1982b: 58-63).

7.4 BRITAIN

The Romanization of Gaul (roughly 120 – 60 BC) moved Britain onto the periphery of the Roman Empire. The Romans first set foot on the island during Caesar's campaigns (55/54 BC) but soon afterwards an outbreak of Gallic revolt diverted Roman attention elsewhere and Rome showed only a little interest in Britain until it was conquered later, in AD 43 (Fitzpatrick & Timby 2002, Millett 1990a: 31, Cunliffe 2000: 119-120). However the two regions influenced each other long before the conquest and the impact of Roman Gaul is seen as a catalyst for the political and economical developments in Britain. Millett (1990a: 31-33) highlights various developments in Gaul that had an impact on the trade with Britain as well as the British Late Iron Age communities: the establishment of peace after the civil wars which provided the basis for administrative centralization and triggered economic development (27 BC) and the construction of a road system which also reached the channel (39-16 BC). Findings of Mediterranean amphorae document existing trade between Rome and Iron Age Britain. The figures suggest that even at this early period Britain was receiving supplies of olive oil as though it was already part of the Empire. The olive oil bearing Dressel 20 accounts for up to half of the total amphorae assemblage at some sites (Williams & Peacock 1983, Peacock 1978). The Gallic tribes of Armorica and the Veneti were in close commercial contact with South-West Britain and when they revolted in 56 BC they received military assistance from the island (Frere 1987: 17) (see figure 10). Even though Caesar's invasion should be seen as grounded in political power plays between him, Pompey and Crassus, rather than economical interest or military necessity, this connection between Gaul and Britain provided a political reason (Frere 1987, Millett 1990a: 40-42).

According to the archaeological evidence at Hengistbury - a port in the area of the Durotriges on the South coast of England (see figure 10) - Italian wine was brought into the country in considerable quantities (Cunliffe 2000: 434-442). Millett (1990a: 30) argues, however, that the quantity of vessels accounts for only a minimum of 720 litres of wine over 50 years and concludes that Britain was only a minor trading partner for Rome. The scarcity of wine could have made it a precious object in Britain with considerable cultural impact. Wine is believed to have provided a new means of displaying prestige which eventually replaced the traditional means of raiding.

The archaeological record shows a change in the pattern of cross-channel trade after Caesar's campaigns (see figure 10). During the pre-Caesarean period the earliest amphorae (Dressel 1A) are mainly found at Hengistbury and in Central Southern England, while in the late first century BC and early century AD finds of Dressel 1B in the South-East dominate. Both vessels contained approximately 20-25 litres of red wine, but were also used for foodstuffs like olives (Fitzpatrick & Timby 2002: 162). While the importance of Hengistbury declined between 50 BC and AD 43, routes on the Rhine made ports in the South-East of Britain (Kent and Essex) more important. Rivers were important means of transportation and trade. In Britain the rivers Severn and Thames provided the main avenues of entry into the heart of the island. They can be divided into two major natural routeways: the south-oriented system along the Severn estuary has Hengistbury Head as the main outlet point to Armorica in Gaul. During most periods both routeways have been used for trade but Sherratt (1996) claims that there is a clear transition from the south-oriented network to the eastern routes oriented on the Thames basin during the 1st century BC. This re-orientation is connected to changes in the settlement distribution, namely the shift from hillforts in Wales and along the Severn to oppida in the South-East.

Caesar's conquest of Gaul and his campaigns in Britain increased the economic and political significance of the South-East coast. Caesar's diplomatic contact with the Catuvellauni and Trinovantes in South-East Britain is thought to be responsible for these developments. The destruction of networks with Armorica has also been named as a possible cause and Cunliffe argues that rather than the result of Caesar's collaboration with eastern tribes these changes are due to the geographical advantage of the Thames estuary in regard to Gaul and the Rhine River (Jones & Mattingly 1990: 57, Cunliffe 2000: 434-442, Fitzpatrick & Timby 2002: 162, Millett 1990a: 31).

The Caesarean period also saw a change in burial practices among the Trinovantes, Catuvellauni, Iceni and Atrebates; while the traditional depositing of weapons in rivers declined, cremations were readopted and became the norm. This has been interpreted as evidence for changes in the socio-cultural system and the adaptation of new beliefs (Cunliffe 2000: 546, Fitzpatrick & Timby 2002).

Like Caesar's campaigns, the reasons for Claudius' invasion of Britain in AD 43 seem to have been prestige-related rather than due to economic incentives. The emperor was in a weak political position and needed military successes to legitimate his power. This hypothesis is supported by the fact that Claudius himself was present when his army took the capital of the Catuvellauni (Millett 1990a: 40-42). It is not certain if the Romans intended to conquer Britain from the beginning or if they simply wanted to end the hegemony of the Catuvellauni/Trinovantes king, Cunobelinus (Jones & Mattingly 1990: 65).

In Britain, the Romans were confronted with a variety of independent political units and it is assumed that they dealt with the tribes by a combination of threat, promises and military action. The tribes that initially presented a problem to Rome were the Durotriges, the Catuvellauni and the Trinovantes (see figure 11). Tribes that remained pro-Roman were rewarded with a form of continued independence and were treated as client kingdoms. The Iceni, the Brigantes, the Atrebates, the Cantiaci and at least part of the Dobunni are generally identified as some of the groups that submitted to Claudius (Millett 1990a: chapter 3). The Roman strategy was therefore more a winning-over of people than a gaining of ground.

There is evidence for maximum utilization of the field systems in the late British Iron Age. This agricultural intensification could have been triggered by the population increase throughout the last millennium BC or by a developing hierarchy in society which left more people out of production but in consumption. Britain is assumed to have shown considerable regional variation throughout this period. While settlement evidence shows that the South and South-East developed a more complex societal organization and the beginning of urbanization, the North and West were less centralized (Millett 1990a: chapter 2). It is assumed that these factors had an impact on the speed with which the Roman army could advance and conquer new territories. The South-East was more developed and urbanized and thus provided infrastructure for the Romans to pick up on, while Wales and Scotland lacked these crucial factors and thus posed serious difficulties. These areas could subsequently only be conquered at a later stage or not at all (Middleton 1979, Cunliffe 1978: 178). It has been assumed that, after the invasion of Britain, products that were essential for the Roman way of living must have increased (Williams & Peacock 1983).

Figure 11: *Iron Age tribes in England and Wales*

7.5 RELATIONS BETWEEN GERMANY AND BRITAIN

Due to the geographical positions of Britain and Germany close contacts between the two countries have naturally been assumed. While this seems true for the armed forces, the presumption of Britain and Germany as natural trading partners has been challenged (Hassall 1978, Greene 1978). Hassall (1978) could show that the sea route between the Rhineland and Britain was used for military purposes. Soldiers from Batavia are likely to have served in Britain from AD 43 onwards. These units must have provided a source for challenging beliefs as they "[...] were not only raised in the Germanies or Gallia Belgica, but also preserved, at any rate for a period, their ethnic character once they had been sent to Britain" (41). In contrast to the large number of German-raised soldiers serving in Britain there is almost no evidence for British soldiers serving on the continent.

While Cunliffe (2000: 543-548) argues that by 10 BC the Rhine estuary became the main route between Late Iron Age Britain and Rome, Peacock (1978) concludes that evidence of the use of the Rhine as a trade route during the 1st to the 2nd century AD means that Britain was not the primary trading partner of Gaul. His argument is based on an economic perspective and the fact that inland waterways were more expensive than sea transport. Of all the existing river routes, the connection Narbonne – Bordeaux is considered to be the cheapest link for a connection to Britain and should have been used to supply the island. The usage of the more costly Rhine route must mean that the Rhineland, rather than the British market, was supplied via this route and Britain was not the primary trade partner. The distribution of Dressel 30 vessels along the Mediterranean coast of France as well as along the Rhine supports this view. However the convenient geographical position of rivers in Gaul made them an integral part of trade and communication (Peacock 1978, Drinkwater 1983: 127).

7.6 OBSERVING CHANGES IN SETTLEMENT PATTERNS FOR THE NORTH-WESTERN PROVINCES

"The spread of civilization could thus be traced in the foundation of cities" (Woolf 1993: 106)

Successful Romanization has been linked to incipient urbanisation and an increase in social complexity. These aspects have been investigated mainly through settlement evidence. A look at the literature reveals that the terminology for settlements is ontologically weak for the Iron Age and Roman period. Such a loose terminology is likely to result in various archaeologists classifying the same object differently. This is significant because changes over time cannot be sensibly monitored if the observation units are not clearly defined. It must be clear that the observed changes are not simply caused by false usage of the terminology: if the use of the terms 'oppida' and 'hillfort' is ambiguous an observed variation in settlement types can reflect behavioural differences as well as a bias caused by differential use of settlement typology. In such cases patterns that look like social change might only conceal social stability. A comparison between Millett's (1990) and Cunliffe's (2000) work on late Iron Age Britain demonstrates the differences in settlement patterns that may arise from a weak typology; Millett (1990a) clearly identifies more settlements as oppida than Cunliffe (2000), who classifies many of the same sites as large nucleated settlements.

Despite arguments that the term 'oppidum' is confusing and imprecise (Woolf 1993, Hill 1995, Trow 1990, Cunliffe 2000: 312) it is still used to describe large proto-urban sites in late Iron Age Europe. The term was first applied by Caesar in his record of the Gallic wars. He uses various terms to describe the tribal centres in Gaul and Britain, talking about oppida (defended strongholds), castella, vici (unfortified centres) and aedifica (farmsteads). His use of the term oppidum includes hillforts as well as oppida and is therefore only of limited use for the separation of the two settlement types in Britain.

For Britain both 'oppida' and 'hillforts' have been used to classify fortified settlements, but the classification is far from unambiguous. Many of the criteria used to differentiate the two settlement types overlap and it thus "seems difficult to sustain the argument that hillforts are a fundamentally different class of site from enclosed oppida" (Hill 1995: 56). For Gaul the use of the term oppidum is also very imprecise. It has been distinguished from hillforts (Collis 1982) but some authors also claim that they are the same type of settlement: "hillforts, termed oppida" (Woolf 1998: 107, Wightman 1985). Oppida in Gaul have also been separated from vici based on their fortification (Nash 1976: 107, Drinkwater 1983: 11). Both settlement types are supposed to be urban and are characterised by a population of which at least part depends on labour other than agricultural occupation. The development of these kinds of settlements was concentrated in Central and Eastern Gaul before the Roman conquest. For Germany there are only a few traces of oppida north of the Rhine.

The following section will introduce the discussion on settlement types.

Existing definitions and separation criteria for hillforts and oppida

Location, date, size, urbanization and fortification are criteria generally used to distinguish hillforts and oppida in Britain. However not all these criteria are generally considered for a classification of settlements; often only two of them have to be satisfied (Woolf 1993: 225). The following section will discuss each of them separately.

Location

As the term 'hillfort' suggests, these settlements are often located on the top of a hill while oppida in Britain are often placed in river valleys (Hill 1995, Cunliffe 1976). However this criterion is not sufficient, as many earthworks classified as hillforts on Ordnance Survey maps are on level ground (Dyer 1992). For Gaul the opposite is the case and oppida are positioned on steep heights and the location supported their defensive nature (Nash 1976: 107). These differences clearly limit international comparison over time.

During the 1st century BC many hillforts in the South-East of England were abandoned and changes in settlement pattern took place which manifest themselves with the appearance of oppida. The hillforts in the South-West and West of Britain were retained and modified with stronger fortifications (Cunliffe 1976, 2000: 366). It has been argued that this shift of location is related to function. Hillforts often occupy interfluvial positions overlooking more than one river valley and oppida are usually lowland centres, sited directly on rivers in Southern England during the first century BC. An increase in economic needs and a decrease in the need for security led to a relocation close to rivers which were significant for trade (Sherratt 1996). This view is supported by Edmondson (1990) who describes how the people of Sabora made a special request to the emperor in 78 AD, asking permission to move their oppidum from its hilltop location to the plain below. The emperor's response shows that this move was considered desirable on economic rather than political grounds. The abandonment of open settlements and their replacement by fortified sites is also common for La Tène Europe (Woolf 1993: 228).

Date

Hillforts were constructed in Britain during the first millennium BC (Dyer 1992, Cunliffe 1976). The majority were constructed between 750 and 400 BC and after 500 BC there is a marked decline in the number of hillforts in constant use. Oppida appear in South-East England during the 1st century BC and are therefore a development of the late Iron Age (Cunliffe 1976). Some were formed out of hillforts which were turned into *enclosed* oppida, like Oldbury and Bigbury in Kent. "The enclosed oppida represent the last stage of 'hillfort' development in south-eastern Britain" (Cunliffe 2000: 368). This is controversial as other authors claimed that many hillforts were not Iron Age Sites at all and the time of their construction varies significantly between neighbouring areas (Hill 1995). In Gaul, oppida first appear in the Late La Tène D around the mid-second century (Nash 1976, Woolf 1993).

Size

Oppida are considered to be large settlements. Cunliffe (1976) classifies them as usually larger than 10 hectares, while Woolf (1995) defines their minimum size as 20 – 25 hectares, but adds at the same time that this criterion is often relaxed, especially in the West of Europe. There is a huge variation in the size of oppida across Europe. While Kelkheim and Manching in Bavaria had rampart systems that enclosed 650 and 350 ha respectively, from the 13 oppida found in Limousin only three were bigger than 10 ha. Woolf (1993) argues that oppida at the margins of their geographical distribution (Hungary and South-East Britain) show an unusual morphology. The oppida lying in between were more uniform but show variance in their rampart construction. For many oppida the 'enclosed' area does not contain any traces of settlement.

Urbanization & function

The discussion on hillforts has been dominated by Cunliffe's redistributing chiefdom model for the Wessex area. Pits and post structures found within the settlements have been interpreted as traces of grain storage. The capacity was beyond both the need and the ability of the people of the site, which led to the idea of hillforts as central storage places with connection to the surrounding hinterland. This system would need some overall control. "Hillforts are essentially a specialised form of settlement: their size, complexity and siting suggests that they represent the communal effort of a large sector of the social group working under the coercive power of the leadership" (Cunliffe 2000: 312). However Hill (1995) has argued that not all contemporary hillforts were the same type of site.

Cunliffe (1976) sees the main difference between these two types of settlement in signs of urbanization found in oppida that are absent in hillforts. While towns were nothing new in continental Europe of the 1st millennium BC, signs of emerging state systems are found in Yugoslavia, Hungary, Czechoslovakia, Southern Germany and South-Eastern Britain with structures defined as oppida. It is often difficult to fully excavate this kind of settlement and most of the excavations have focused on ramparts and entrances (Cunliffe 2000: 312). Knowledge about the functional usage is therefore limited. The presence of coin flan moulds, imported glass, ceramics and foodstuffs which quantitatively and qualitatively outnumber finds in surrounding non-nucleated settlements is seen as evidence for early urbanisation and common to several oppida sites (Trow 1990).

However other authors have argued for signs of urbanization in hillforts, also. Evidence of trade, industry and coinage also exist at many late hillforts, for instance Maiden Castle or the Wrekin (Jones & Mattingly 1997: 47). The occurrence of such artefacts alone is therefore not an adequate classification criterion as the transition is gradual and qualitative as the following quotation shows: "The processes of urbanization are *even more dramatically apparent* in areas where hillfort occupation had declined virtually to the point of non-existence by the end of the first century BC." (Jones & Mattingly 1997: 47). This quotation claims that there is a difference, but that his difference is qualitative rather than quantitative. For Gaul, urban settlements have been distinguished from other types of settlement through their dependency for at least some parts of their livelihood on activity other than agriculture (Nash 1976, Drinkwater 1983: 136). Nash (1976) sees the aspects of trade and administrative function as main criteria for oppida in Gaul. The oppida of the Treveri are seen as centres of craft production and trade (Carroll 2001: 21).

Fortification

Hillforts are classified according to the number of ramparts and ditches surrounding them. *Univallate* describes hillforts with only one rampart and *multivallate* those with more than one (Dyer 1992, Cunliffe 2000: 322). Multivallate hillforts are generally a later development and the extra ditches and ramparts are thought to have been added for defensive reasons. When the number of early hillforts declined in Southern Britain, the hillforts that survived in the West were modified and extra fortifications were added. This type is classified as developed hillforts. These developments are seen as evidence of an increased concern for strong defences. Examples in Britain are Maiden Castle, Danebury and Hambledon Hill (Cunliffe 1976). For Gaul, towns have been distinguished from agricultural villages. Towns were divided into non- fortified vici and fortified oppida (Nash 1976). The defensive position – often on elevated ground – is an important criterion for oppida.

7.7 CONCLUSIONS

There is evidence that the expansion of the Roman Empire in all three regions depended on the existing pre-Roman structures and social organization. The successes of Roman campaigns are related to the level of social development in the new territories, and that has been linked to the level of urbanisation. A comparison of areas where the Roman campaigns went smoothly and successfully with areas that posed more difficulties confirms these assumptions. While Gaul had undergone processes towards an urban-centred state and the Romans could conquer this area relatively quickly and effectively, the dispersed settlements and single independent tribes in Germany and Northern Britain provided far more difficulties (Cunliffe 1978: 178). In Central and Eastern Gaul the Romans adopted the existing administrative borders, as the development in these areas had reached a level that provided a basis for the Romans to work on. Central and East Gaul was the "[...] area within which Gaul first developed the political institutions of the archaic state, and it was the area of Gaul which most readily assimilated Roman social and political institutions after the Caesarian conquest" (Nash 1976: 95). In Brittany, the Alps, Normandy and the North of France where the archaeological evidence suggests barbarian persistence until the conquest, urban settlements appear to be far less developed (Nash 1976: 107 f).

In his records of the Gallic wars Caesar makes a similar connection between urban centres and the degree of civilization in the Celtic tribes. He divides the Celtic tribes in Britain and Gaul into barbaric and civilized and argues that the proximity to Gallic culture is responsible for the high grade of civilization: "they live beside the Rhine and traders come to them frequently, and because they are themselves close enough to the Gauls who have become accustomed to their way of life" (4:3). He identifies the people of Kent as by far the most civilized in Britain claiming that "their way of life is much the same as that of the Gauls" (5:14). Even though we should consider Caesar's - clearly politically motivated - categorisations carefully, the citations confirm that similarity to the Roman culture was an important criterion for smooth Romanization.

Archaeologists have argued that the absence of oppida may reflect an underdeveloped socio-political hierarchy. This argument ties living in cities to the concept of civilization (Nash 1976, Carroll 2001: 21, Woolf 1998: 106). Settlement patterns would therefore provide a good means for observing social change, however the distinction between hillfort and oppidum is far from clear. This chapter has demonstrated that both terms describe very heterogeneous objects, which vary in size, complexity and location. In the case of hillforts it also describes a settlement form spanning 900 years.

Such difficulties with the classification of phenomena are not unusual for archaeology. Categories like 'urbanised' and 'pre-urbanised' or 'agricultural' and 'pre-agricultural' are often fuzzy, because they try to pin down multi-dimensional phenomena. The number of dimensions by which an object is defined influences the degree of certainty with which this can be achieved (van der Leuw 1989). A definition of settlement types using only one of the six criteria discussed above would allow a far more unambiguous separation of sites. "It is possible to distinguish and define clearly in any one of a number of dimensions, but impossible to keep clarity when more than one dimension is involved at the time" (ibid: 309). Thus – paradoxically - an increase in information often results in the loss of clarity. This shows why modelling is such a valuable tool, because it requires exact decisions about all the dimensions considered in the classification. However, while the goal of this research is not to create a typology for Iron Age settlements, the fuzziness of the widespread classification should be considered when interpreting the archaeological record.

8. THE ANALYSIS OF FAUNAL REMAINS: QUANTIFICATON AND PREVIOUS APPROACHES

Both the literature review on the North-Western Provinces and the theoretical model on innovation suggest that significant changes appeared *before* the Romans actually occupied new areas. The first part of this book has discussed the issue of innovation cascades, where one innovation creates fertile circumstances for further changes until the old conditions have almost vanished. Major cultural changes like Romanization are likely to involve a number of smaller innovations that interact and accumulate in this fashion. By the time such behavioural changes become visible in the archaeological record, the actual innovation is already over.

The next part will look at faunal remains across Iron Age Britain to see whether there are significant differences between certain areas: the hypothesis assumes that the South-East of England, which could be conquered by the Romans with relative ease, must differ from Scotland and Wales in as much as incipient changes within the Iron Age societies created conditions which made it accessible for the occupiers. The literature review has shown that there are a number of changes in Time-Space constraints that support this idea. The analysis of faunal remains will be used as evidence for changes in behaviour manifested in the archaeological record and the results will then be connected to the evidence of changes in Time-Space constraints.

Faunal remains have been used to gain insight into past economies and subsistence strategies in various ways; the relative proportion of one species in the overall assemblage gives evidence about its role in the original economy. An investigation into the relative proportion of each skeletal element can give insights into the processes that have influenced the sample before and after it entered the archaeological record. The evaluation of archaeological artefacts from different sites and time periods is only possible if the data are recorded in a format that allows a direct comparison. The same is true for intra-site comparison or investigations about inter-species variation within one context. The next section will introduce various quantification methods (NISP, MNI and GUI) that are used to achieve such comparability for faunal remains. The choice of quantification method can have an impact on the representation of the original pattern. These potential biases will also be considered.

8.1 QUANTIFICATION OF FAUNAL REMAINS: NISP, MNI & GUI

NISP values give the "number of identified specimens per species" and are the most commonly used unit of quantification. A NISP count includes both complete bones and fragments and can be expressed either as the overall number of finds per species or in more detail for each individual skeletal element per species. This method has a number of disadvantages; the species considered within a study may have different numbers of bones in their skeletons; pigs for instance have more foot bones than sheep. Species with more bones have potentially higher NISP counts and may therefore be over-represented in an inter-species comparison. Also fragmented bones will contribute more to a NISP count than unfragmented elements. Fragmentation results in the same bone being counted several times unless all fragments from one element are put together and counted as one, which is difficult in practice. Experience has shown that different species recovered from the same context were affected differently by taphonomic processes (Maltby 1987, 1995). The less well preserved species may be more fragmented (raising NISP) but are also more likely to be absent altogether (lowering NISP). We cannot know how these factors are resolved in any one study.

Thus taphonomic processes, which can be inferred in literally every archaeological assemblage, may artificially increase the NISP frequencies. This is not only problematic within one site but may especially affect inter-site comparisons. Finally NISP counts do not work when complete or partially articulated skeletons are present, which is not unusual for the British Iron Age where such finds occur as 'special deposits' within ritual sites. Most mammals have well over 200 bones and one or two complete burials may raise the NISP values of that species dramatically, especially when the overall assemblage is relatively small. In this case the whole skeleton can be treated as a single specimen or be excluded from the NISP counts (Chaplin 1971: chapter 4, Klein 1980, Hambleton 1999: chapter 5).

MNI or the "number of individuals which are necessary to account for all of the skeletal elements (specimens) found in the site" (Shotwell 1955: 327) can be calculated in various ways. In a first step the minimum number of examples of each bone is determined. Because some bones are individual (mandible) and others paired (long bones) or multiple (vertebrae), there are different methods to calculate MNI. The number of multiple bones is generally divided by the number of them normally present in each animal, while the MNI for paired bones can be determined in different ways. First the bones are divided into left and right hand side of the body. In the crudest application of this method the most abundant body side gives the minimum number of individuals by which the species as a whole is represented in the assemblage for that skeletal element. The existence of 9 left side humeri and 5 right side humeri for example means that at least 9 animals were deposited at the site.

In a more detailed analysis age and size of an animal are considered as well and the MNI is calculated as follows:

$$L + R - P$$

L: left hand side bones, R: right hand side bones & P: number of pairs.

By comparing the size and age of the bones they are paired and that number is subtracted from the sum of left and right side bones. For instance if all the 9 left side humeri are from large and adult animals and 3 of the 5 left side humeri are from juvenile animals while the other two are from large adult specimens, there must have been a minimum of 12 animals to account for the recovered humeri (9 + 5 – 2) (Chaplin 1971: chapter 4, Ringrose 1993).

This quantification method is influenced by the way in which the material is grouped before the minimum numbers are determined. It depends on whether archaeological layers are separated or the data is grouped according to age, sex or body sites. Grayson (1973, 1978) describes the detailed grouping of faunal remains as the *maximum distinction approach*. If the left and right hand side bones of the skeleton are separated and the most abundant number of both sides are considered the MNI counts are generally higher than if the bones are counted without this division and the result is divided by two. For instance if 24 pig femora are found, the ungrouped MNI would be 12 calculated by dividing the total number of recovered pig femora by the number of femora in one pig. However if body side is considered, 20 femora could be right and 4 left, which would result in a MNI of 20 rather than 12. Grayson has shown that the higher the number of clusters, the higher the values for the MNIs. This means that the MNI values will be smallest when the data is recorded with the so-called *minimum distinction approach* which treats the material from one site as one single large cluster (Chaplin 1971: chapter 4, Grayson 1973, 1978, Klein 1980, Speth 1983: chapter 3).

This phenomenon is related to another potential bias caused by MNI: the impact of sample size. This quantification method tends to exaggerate the importance of rarer animals, because single elements within smaller assemblages make proportionally greater contributions to the MNI values than do single elements within larger bone samples. The maximum distinction approach creates smaller sample sizes and thus exaggerates the bias caused by smaller samples. Grayson (1978) has shown that very large samples might have the same distorting effect and the value of MNIs can be minimized when very large samples are compared to smaller ones.

Based on MNI counts, Binford (1978) has developed a quantitative measure of bone frequencies that can account for the impact of human activity on the composition of bone assemblages. Based on an anthropological study of Numamiut Inuit hunting strategies Binford created the "general utility index" (GUI), which considers the 'usefulness' of each skeletal element. Binford argues that meat is generally not transported, shared, stored or cooked as whole animals and we should therefore not expect to find whole skeletons within the sites of hunting populations. Furthermore the meat, marrow and grease content of each skeletal element influences the way hunters treat each part of the carcass and thus the likelihood of various parts entering the archaeological record. Binford therefore weights each skeletal element based on these factors. For sheep the GUI of ribs, sternum, distal femur and pelvis are highest. However Binford points out that this index is accurate but impractical. Animals are generally not butchered into single bones but rather articulated low utility elements like phalanges are grouped with high utility elements. The GUI was thus adjusted into a "modified general utility index" (MGUI) which takes the consequences of transportation or butchery practices into account (Binford 1978: 69-75, 1981, Speth 1983: chapter 5).

While Iron Age and Roman societies were farming rather than hunting groups, the amount of calories that can be gained from various skeletal elements should still be expected to have an impact on the way the carcasses were treated. For cattle the scapula, humerus, radius, ulna, sacrum, pelvis, femur and tibia are the highest meat-bearing elements, whereas the MGUI for sheep is highest for ribs, sternum, pelvis, distal and proximal femur.

8.2 POTENTIAL INFLUENCES ON THE FREQUENCIES OF FAUNAL ASSAMBLAGES

Biases of the original data can already occur before the quantification of the archaeological remains. Each artefact that is recovered from the archaeological record had a life history during which it was part of a complex cultural system (Binford 1981: 87). The frequencies of skeletal elements recovered within an archaeological context almost always differ from the expected frequencies of complete skeletons (Brain 1981: chapter 2, Klein 1980). The deviation from such an unbiased distribution is caused by various processes that acted on the bone before and after it entered the archaeological record as well as while it was in the ground.

We can crudely differentiate three types of influences that act on animal bones: 1) processes acting on the bones *before* they enter the archaeological record, 2) *while* they are in the ground and 3) processes that influence their history *during and after recovery*. The first group mainly summarizes influences of bygone human behaviour as well as other animals (butchery practices, animal gnawing, cooking, how the bones get into the ground), the second group includes all the taphonomic processes acting on the bone while it is in the ground (microbial attack, soil acidity, density of individual bones). Both lie beyond the control of archaeologists. Only the last group of factors can be influenced by the archaeologist (differential excavation techniques, publication choices).

The identification of the processes that led to a particular distribution is complicated through the fact that all three bias groups might alter the material simultaneously and in similar ways. Each distribution of faunal remains is likely to be the result of various interrelated as well as independent processes. Any quantitative intra- or inter-site comparison has to consider these three potential sources of bias and it is very likely that all three processes had an impact on the faunal remains of the British Iron Age. The reconstruction and identification of such processes is an essential part of the data interpretation and might reveal valuable insight into its background: in order to understand the composition of an assemblage it is necessary to understand the history which might have led to its specific features.

Potential biases before the bones enter the ground

Anthropological case studies and experimental archaeology have been used to gain a better understanding of the first group of factors that influence the constitution of faunal remains. It can be shown that humans are not the only species that might manipulate bones. Animal scavenging and gnawing often influence animal remains that were not deposited into the ground straight away. These processes might leave traces that are inseparable from human impact (Brain 1981: chapter 2, Binford 1981: chapter 5). Binford's comparative study of various Indian groups has shown which parts of the skeleton are likely to be removed prior to transportation. Based on this knowledge, variation in the frequencies of certain parts of the skeleton can be interpreted as the result of a particular behaviour and give information about the use of that particular site (1981: chapter 4).

Potential biases while the bones are in the ground

A basic understanding of the constitution and development of bones and teeth is essential for the correct interpretation of the pattern caused by taphonomic processes. Bones and teeth are the organic materials with the highest survival rates in the archaeological record. This is due to their composition which gives them exceptional robustness. Bone is made out of water, minerals and organic material. The organic material is responsible for its elasticity and 90% of it is collagen, a very robust protein that cannot be dissolved in natural conditions. The inorganic phase gives a bone its toughness and mainly consists of the mineral Hydroxyapatite $Ca_5((PO_4)OH)$ which accounts for 65% of the bone weight. The relationship between these two components changes throughout life and determines the weight and elasticity of a bone. While a newborn has about 50% mineral in its bones an old individual has about 70% and this increase in mineral content makes the bone less elastic and easier to break (Platzer 1991: 14-17, Davis 1995: chapter 2).

Once the bone has been buried in the ground there are various factors that may influence its survival rate. Most carcasses are attacked by bacteria very quickly after death and this might result in the destruction of organic material. Flesh provides ideal conditions for various micro-organisms and about 75% of all bones found in an archaeological context show traces of such attacks. It has been shown that in human burials the thorax bones are more severally attacked than the long bones, which is due to the high numbers of micro-organisms in the gut (Child 1995). Animal bones often show less microbial attack than human bones, because their carcasses are generally used for meat, which means that bones are mostly clean of soft parts before they enter the ground. Acid soils tend to dissolve away the bone mineral which decreases the survival rate of the bone in so far as only the most resistant parts of bones and teeth will survive. Soils with high rates of water flow also lower the survival rates of bones significantly (Chaplin 1971: 13-19).

Based on its morphology, bone is divided into different groups: *long bones* which are confined to the limbs, *flattened bones* like the pelvis, scapula or parts of the skull which have large surface areas for the attachment of muscles, *compact bones* like the tarsals and carpals and *irregular bones*, like vertebrae which combine various elements of the other groups and can therefore not be classified into any of them. These bone types vary in their relationships of compact and spongy (cancellous) bone, which determine the strength and weight of a skeletal element and are influenced by function and load. Bone structure is a trade-off between maximum strength and minimum weight. A higher proportion of the more robust compact bone increases the survival rate of a bone, because it is less affected by taphonomic processes. Bones also differ in the way they ossify. Long bones have three primary centres of growth: the shaft and the two ends, the so-called epiphyses. Once the bone growth is completed these epiphyses are closed and united with the shaft (Chaplin 1971: 13-19, Davis 1995: chapter 2, Platzer 1991: 14-17).

Brain has combined both factors and argued that they have a major impact on the differential preservation of different parts of the skeleton. The toughness of a bone is determined by the relationship of compact to spongy bone and the closure of its epiphyses. Both factors may vary, which leads to differential intra-bone preservation. The distal end of the goat humerus for instance is composed of compact bone and has an early fusion time while the proximal end is wide, thin-walled and filled with spongy bone. For the radius, femur and metatarsals the proximal end is expected to outlast the distal end. Teeth are generally the most durable body parts and the mandible is a very robust bone and easily identifiable, even if fragmented. It is therefore not surprising that Brain's study on goat bones shows that mandibles and distal humeri have the highest survival rate, while only very few vertebrae, phalanges and distal femora could be recovered (Brain 1981: chapter 2).

The robustness of a bone also determines its likelihood of fragmentation. Elements with a high component of spongy bone are more likely to break. The previous section has discussed the potential biases that fragmentation can cause in the bone counts of a site. The form of a bone also plays a part in this; some bones are easier to classify when fragmented, because of their characteristic and distinctive shape. On the contrary, fragments of bones with similar shapes to others, like the long bones, are less likely to be identified and have a higher chance of being under-represented in the bone report. Ribs and vertebrae are also often badly preserved because they are fragile and in many cases their fragments cannot be classified.

NISP counts can be used to determine the degree of fragmentation in the assemblage. The division of NISP through the bone weight gives an indication of the average weight per bone fragment. High NISP counts and low bone weight will result in a high value, indicating small bone pieces and thus high fragmentation.

Strong heat results in the combustion of the organic components of a bone including the collagen. As a consequence it becomes brittle and has less chance of surviving the archaeological record. Thus cooking or roasting reduces the chances of a bone surviving in the archaeological record. Some bones are more likely to be cooked than others, which depends mainly on their meat content.

Potential biases after the artefacts are recovered: recovery techniques & confounding

The last group of factors that determine the composition of faunal remains is the only one that is influenced by the archaeologists. The last decades have seen significant developments in excavation and documentation techniques which are partly connected to a rising awareness of the impact archaeological decisions can have on our understanding of the past. Payne (1972a) has demonstrated the impact of recovery techniques on pottery, bone and stone. He could show that the excavators frequently miss parts of the original assemblage and that the recovered sample is often not a valid representation of the original assemblage but heavily biased. Different excavation techniques like sieving can significantly alter the relative abundance of bones, including for larger species like cow, pig and sheep, where even the smallest bones are of a size that should not be overlooked easily.

Archaeological remains recovered at a site often come from different contexts (pits, ditches etc.) and phases. Archaeologists generally interpret patterns of artefacts *between* time phases as time-trends and patterns *within* one phase as the result of behavioural specialisation. While these assumptions are reasonable for most cases, they might be falsified for some data, due to the so-called '**confounding problem**'. This phenomenon might arise when categories of data are collapsed and is also known as the Simpson's paradox (Shapiro 1982, Winder pers. comm.). Grouping data from one site is problematic even for contexts within close spatial proximity because the samples might have very different histories of use, disposal or post-depositional processes. The grouping of data from different phases might also cover important intra-phase variation. One phase, for instance, might have more pits than another and the variation between layers might be due to differences in recovery contexts rather than time.

Confounding describes the fact that the collapsing of categories might create false, or overwrite existing, time trends in the archaeological data. This will be demonstrated with two fictitious examples. Table 1 shows how the collapsing of categories can *overwrite* an existing time-trend. While type 1 pots are more abundant in the ditches of phase 1, type 2 pots are more abundant in the pits. This pattern is reversed for phase 2 where type 2 pots are more abundant in the ditches and less abundant in the pits. The pottery data shows a clear time trend with more type 2 pots entering the archaeological record through ditches in the later phases while they become less abundant in pits. This could indicate changes of behaviour at the site. However this time trend disappears when the pots recovered from ditches and pits are treated as one large sample. The cross-classified data only shows an insignificant increase in type 2 pots.

	Ditch		**Pit**		**Ditch & Pit**	
	Pot 1	**Pot 2**	**Pot 1**	**Pot 2**	**Pot 1**	**Pot 2**
Phase 1	60	20	40	80	100	100
Phase 2	50	100	30	10	80	110

Table 1: *Collapsing categories can overwrite an existing time trend (altered after Shapiro 1982)*

Table 2 shows that collapsing cross-classified data can also *create* false time trends. Looking at the data separated for phases and context, it is obvious that type 1 pots are more abundant in the ditches, while the type 2 pots are more abundant in the pits for both layers. The observed pattern remains the same throughout the time of observation. However if both contexts are collapsed the data begins to show a clear time trend; type 1 pots are less abundant in phase 2 while there is an increase of type 2 pots in phase 2.

	Phase 1		**Phase 2**	
	Pot 1	**Pot 2**	**Pot 1**	**Pot 2**
Ditch	120	60	40	20
Pit	20	40	50	100
Ditch & Pit	140	100	90	120

Table 2: *Collapsing categories can create a false time trend (after Winder pers comm.)*

The collapsing of categories created during an excavation can therefore have a significant impact on the pattern that archaeologists observe. However the original data often contains very small sample sizes which make statistically valid statements about the composition of the entire population impossible. The aggregation of several small sites or various contexts from a single site into one large sample is often the only way to make the available data more robust and suitable for statistical analysis. Thus archaeologists are facing a dilemma: the nature of the archaeological record forces them to collapse categories in order to make it statistically robust but at the same time this may lead to confounding and the creation of false trends.

The data that will be used to test the conceptual model (see section 9.1) has also been grouped; each site is represented by one assemblage of animal bone frequencies regardless of the more detailed recovery contexts and sometimes even temporal layers. It will therefore be impossible to exclude confounding biases from the pattern that will be observed. However this is accepted because it is seen as a necessary trade-off to get quantified data and test the hypothesis.

8.3 PREVIOUS STUDIES OF CULTURAL CHANGE BASED ON FAUNAL REMAINS

A number of studies have investigated the transition from Iron Age to the Roman period based on faunal remains. Seeman (1987) suggested using faunal remains to trace Roman influences on the indigenous population. He argued that better breeding methods introduced by the Romans would have increased the size of cattle which could be observed through time with metric analysis of skeletal elements. Furthermore trade between Romans and indigenous groups could have altered the species distribution at Romanized sites. The Roman troops are thought to have consumed high quantities of young pork and an adaptation to their demands is likely to have an impact on the age and gender distribution as well as the relative number of pigs within a settlement. These hypotheses were tested on 4 sites in the Assendelver Polders in the Netherlands by comparing the MNI of the faunal remains of cattle, sheep/goat and pig for time periods before and after the Roman conquest. Due to poor preservation in acid soil the number of skeletal elements proved too small for any intra-side comparison. The data did not allow the observation of any Roman influence and could only conclude that cattle, sheep/goat, pigs and horses were consumed.

By using literature sources (writings and pictorial representation) as well as faunal remains Peters (1998) could show that Roman husbandry regimes had an impact on the animal production in the Germanic parts of the North-Western Provinces. Metric studies suggest that new breeding methods came into use soon after the occupation of new areas. Peters argues that horses became bigger to meet the needs of the Roman army which considered the Germanic and Celtic species as unsuitable. Cattle were also bred to increase their size which made them more suitable for agriculture and transport. The Romans also tried to increase the meat and fat content of pig by increasing the average body weight of the animals and slaughtering them relatively late at approximately two years. Of the three main domesticates only the sheep shows no visible impact of Roman husbandry regimes.

A study of the Mosel Area in Germany has analysed the faunal remains of 15 sites dating from the Iron Age to the Roman period. First results show that while the importance of the main domesticates, cattle, sheep/goat and pig, remain basically unchanged there are a number of alterations through time. For the Roman Period the appearance of wild animals like pheasant and peacock that had to be imported over long distances could be shown. In addition cats and pigeon bones appeared with the Romans and an increase in poultry as well as an intensification in game hunting could also be observed (Oelschlägel 2006).

The Archaeomedes project studied settlement pattern rather than faunal remains, but the analysis provides an excellent example for the observation of change through the use of quantitative, statistical analysis on archaeological remains. The study investigated 934 sites in 8 sample zones in the Rhône Valley, covering the period from 120 BC to AD 500. This covers the whole period of Roman impact on the area. In order to observe changes in settlement pattern the 44 parameters describing the settlements discovered in the area were clustered statistically by factor analysis. This statistical creation of observable units provides independence from the classification found in the archaeological literature. The previous chapter has discussed the inadequacy of settlement classification for the Iron Age and has argued that important aspects of change can be overwritten by such an imprecise usage of terminology. The statistical clustering of settlement types, however, can avoid this downfall because the settlement units are based on actual similarities of crucial factors. The project could demonstrate a rapid acceleration in the colonisation of the Rhône Valley during the first two centuries of the Roman occupation (100 BC to AD 100); 80% of the new settlements were erected during these early stages. The development starts in the South of the research area in the Alpilles, the Vaunage and the Uzège. The exception is Lunellois where hardly any settlements were built between 50 BC and AD 50; this has been linked to the lack of pre-existing urban infrastructure in this hilly area. In accordance to existing ideas on Romanization this study shows that in the earliest phase of occupation the Romans chose areas for the creation of new settlements that already had pre-existing infrastructure (van der Leeuw et al 1998, Favory et al 2003).

For Britain there are two main comparative studies of the chief domesticates, cattle, sheep/ goat and pig: Hambleton (1999) for the Iron Age and King (1984, 1991, 1999) for the Roman Period. Both use NISP values and visualize the results with tripolar graphs. The comparison of NISP percentages has revealed that the Romanization of Britain resulted in a shift from high sheep percentages found at Iron Age sites to high frequencies for cattle in Romanized locations, while non-Romanized sites continued to show a stronger emphasis on sheep husbandry (King 1984, 1999, Hambleton 1999). Thus the species distribution seems to depend on the degree of 'Romanization' of a site. King (1984, 1999) argues that a high percentage of cattle in the overall assemblages represents the diet of the Roman Army, whose presence in Britain led to changes in the diet of the indigenous groups.

These results suggest that the data might reveal higher frequencies for cattle bones in assemblages that date into the Roman period of Britain, while the earlier phases are more likely to contain higher proportions of sheep.

8.4 CONCLUSIONS

This chapter has introduced a number of quantification methods that allow the comparison of faunal remains. It has demonstrated how the type of quantification can influence the assessment of the relative abundance of the species. Generally MNI is considered to be the most suitable approach for comparative studies because it is less affected by recovery biases and has fewer distorting effects than other methods (Payne 1972b, Chaplin 1971: 68-75). The chapter has also introduced three groups of factors that influence the history of bones, from the slaughter of the animals in antiquity, their incorporation into the archaeological record, to their documentation and interpretation by archaeologists.

A number of studies have demonstrated how faunal remains can be used to study the process of Romanization. For Britain there appears to be a change in subsistence strategy with an increase of cattle in the Roman period.

9. INTERPRETATIVE DATA ANALYSIS: PART I

After a brief summary of changes in Time-Space constraints in the research area, this chapter will search for evidence of change of behaviour that can then be linked back to the conceptual model of innovation. A first data analysis will apply visual techniques (histograms, scatter plots, box-and-whisker plots) to faunal data from the British Iron Age in order to identify patterns. The dataset will be approached from many different angles (for instance both inter- and intra-site comparisons) in order to get a broad understanding of the information.

9.1 THE DATA SET

This chapter uses data of published faunal remains from the Iron Age in areas that later became the North-Western Provinces of the Roman Empire. The data was obtained from Hambleton's (1999) study on Animal Husbandry Regimes in Iron Age Britain, which includes frequencies for the three main domesticates: cow, sheep and pig. Her study summarizes all suitable published data for sites excavated after 1950 that were classified as settlements. When possible, data from ritual centres were excluded but this distinction is not always easy to make. The Danebury hillfort, for instance, has areas of ritual significance within the settlement. The dataset distinguishes various site characteristics: six regional groups broadly based on shared cultural and geographical traits; four time periods from early Iron Age to Late Iron Age and Romano-British; four settlement types including banjos, hillforts and open as well as enclosed settlements and data on geography and topographical location (Hambleton 1999: chapter 3).

The settlement information might already point towards changes in Time-Space constraints but bears the danger of being very imprecise due to a very loose typology for the British Iron Age (see section 7.6). Oppida, which have been ascribed an essential role in Romanization (Cunliffe 2000: chapter 14) are not identified as an individual group but rather clustered with various other settlement forms in the category 'open settlement'. The crucial characteristic for settlement types in this study seems to be the topographical location; hillforts are generally located higher than 76m OD while oppida are lowland settlements. None of the 'typical' territorial oppida of South East Britain, such as Colchester and Silchester, yielded faunal data that could be included in the database (Hambleton 1999: 17).

The data published in appendix 3 gave the NISP values for the skeletal elements of 46 sites throughout Britain. These frequencies provided the information for the database which was created in EXCEL 2000 (Microsoft 1999) (see appendix 1). Only the site reports for Danebury and Mount Batten gave two frequencies for most bones, one for the distal and the other for the proximal end. In order to adjust them to the other site reports the higher number is chosen to represent the bone. Theoretically each proximal end could belong to a distal end; therefore the smaller number is ignored rather than adding up both values, which carries the danger of counting the same bone twice. This might be crude but is considered to be the best method of including these assemblages in the analysis.

Although MNI data are less likely to be biased by differential preservation, the lack of Iron Age faunal data quantified with this method means that this case study had to rely on NISP frequencies.

9.2 EXPLORATORY DATA ANALYSIS OF BRITISH IRON AGE FAUNAL REMAINS

The sites included in this study are very likely to vary in size as well as quality of excavation and documentation. It is thus unlikely that the dataset possesses statistical properties that allow easy description or prediction. There will be few normal distributions for example. *Exploratory Data Analysis* (EDA), rather than classical statistics, is chosen to determine the nature of the Iron Age remains. Rather than being a set of unique methods EDA is more of a particular way of approaching the data. This technique was developed by Tukey in 1977 and stresses the importance of methods that visualise the dataset, because it claims that they will maximize the information that can be gained from the data. This has two reasons, first summary statistics of location and spread may give only an incomplete picture of important data characteristics and, second, the shape of a data distribution is best perceived visually and there are no quantitative analogues that will provide the same insight into the data. EDA stresses the principle of *openness* in as much as the researcher should be looking for possibilities that he or she did not expect to find and reject any attempts to fit the data into an anticipated pattern. Graphical techniques are ideal because they allow an exploration of the data without any presumption about its underlying structure. EDA does not reject statistics but argues that it should not be relied upon exclusively: "Exploratory data analysis can never serve the whole story, but nothing else can serve as the foundation stone, the first step" (Tukey 1977: 3, Hartwig & Dearing 1979, Toit et al 1986).

A primary analysis is undertaken by plotting the frequencies of species and skeletal elements in scatter plots and histograms as well as box-and-whisker plots. The relationship between variables has three main characteristics: shape, strength and direction. The description of these characteristics is generally the most important step in any data analysis.

Scatter plots are very effective because they provide information about all three characteristics. They can be used to give a visual impression of the relationship between two quantitative variables before they are described with statistical methods. The *direction* of the data is described by the spread of the observed values; if they plot from the bottom left to the top right corner the relationship is positive. In such a case high values on one variable are associated with high values in the other. If the plot ranges from the top left to the bottom right corner the relationship is negative and high values on one variable are associated with low values in the other. The *shape* of the dataset is indicated by the form of a line that can be approximated through the observed values. The *strength* of the relationship between the variables is specified by the extent to which the data points conform to the shape: the greater the spread around the line the weaker the relationship. Furthermore, scatter plots allow the identification of outliers, values that do not fall within the overall pattern. Such outliers may be crucial because they can distort statistical summaries of the dataset (Hartwig & Dearing 1979: chapter 3).

Another very common way to visually represent the *profile* or composition of the data is the use of *histograms*. They are mainly used for categorical data, but can also be used to visualize the distribution of quantitative variables when they are clustered into groups. They provide graphical information about the centre and spread of the data as well as the presence of multiple modes or outliers. Usually the data is split into equal sized bins or classes and the number of points from the data that fall into each class are counted. The frequencies are represented on the vertical axis. The peak of a distribution gives the mode or value that occurs most frequently and a normally distributed dataset is the classic example for a unimodal distribution (Benninghaus 1998: 30-36). All three methods are simple but very effective.

Box-and-whisker plots were developed by Tukey (1977: 39-41) and represent a dataset visually with 5 statistical parameters. As the name indicates, the plot contains a box with 2 whiskers. The line within the box represents the median of the distribution; the outsides of the box (hinges) give the first and third quartile and thus identify the middle 50% of the data. The ends of the whiskers give the lowest and highest values. Such plots are useful for depicting the location (centre), spread (scale) and skewness of a data set. In a completely symmetrical distribution the median divides the box into two equal halves and the whiskers are of equal lengths.

This chapter will give a summary of the basic characteristics of the dataset and note any observable pattern and outliers. Several methods will be used to search for differences and similarities in species proportion as well as the abundance of skeletal elements within and between the Iron Age assemblages.

The greatest challenge will be to differentiate between inconsistent patterns that were caused by behaviour and might thus indicate innovation and such outliers that mirror the distorting effect of taphonomy or excavation techniques. All graphs were prepared in Genstat® for Windows (1995) and/or EXCEL (Microsoft 1999).

9.3 TWO VARIABLES TO REPRESENT SPECIES FREQUENICES

In the original dataset two different measures for the overall frequencies of each species are given for each site; the dataset contains the frequencies of all the bones that could be allocated to one of the three species (TOTAL NISP values) as well as all the number of cases for which both species and skeletal element could be identified (SUM NISP values). Both variables describe the *scale* or size of an assemblage. While the latter provides more detailed information, the sample size of the TOTAL NISP values tends to be bigger, because not every skeletal element might have been individually listed in the bone report but added to the overall list. Both variables therefore differ in completeness and detail and the plots in this chapter are also used to investigate whether the results differ in regard to which size variable has been used and if these differences are significant. The two values could not be compared for all sites, because in some cases the labelling was not accurate enough. The total NISP numbers for Owslebury, for instance, are separated into four time periods, while the skeletal element count only names the site without allocating it to a specific time period. In order to prevent a mix-up only the calculated SUM NISP values are considered for these cases. Therefore only 35 out of the original 46 sites could be compared.

Figure 12 shows the relationship between the two scale variables for all three species. Visually all the data points should lie on or beneath a straight line drawn through the scatter plot, because the SUM NISP values must logically be lower than or equal to the TOTAL NISP value. This is the case for all but one site. For Burgh LIA the SUM NISP value of 2021 is higher than the TOTAL NISP value of 1460. This difference can be explained by the fact that the skeletal element data for Burgh includes 'cow size' and 'sheep size' fragments while the total NISP counts only include those fragments that are definitely identified at the species level (Hambleton pers. comm.). Sheep size fragments can also contain elements from other species, like goat, pig, roe deer or dog (Maltby 1987). This assemblage should therefore be considered carefully before any further analyses are undertaken.

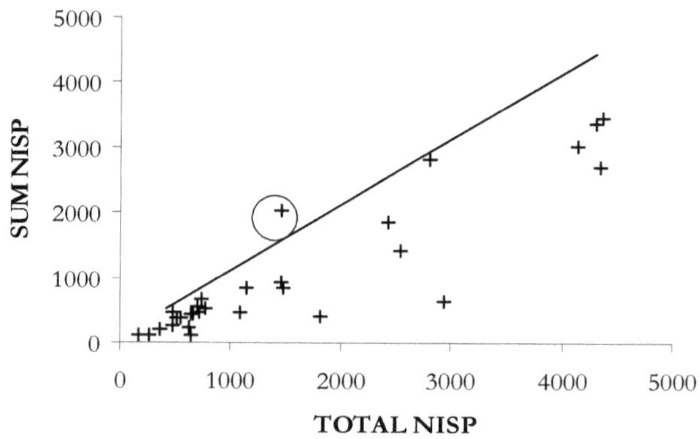

***Figure 12**: Relationship between TOTAL NISP and SUM NISP for 35 Iron Age sites. The line indicates the points at which the TOTAL NISP frequencies are equal to the SUM NISP values*

9.4 DATA ANALYSIS AND SCALE

The two graphs show that the difference between both overall frequency counts is higher for the largest 5 assemblages, for which significantly more bones have been assigned to a species but not to a skeletal element. This could lead to a distortion favouring the large assemblages when the species distributions are compared. Therefore comparisons are made for both variables to investigate the dimension and nature of their influence.

A one-sided analysis of a complex phenomenon like behavioural change can only provide limited insight. Therefore various scales should be simultaneously investigated and compared. An intra-site comparison of the artefact distribution, for instance, can provide information about the use of this individual site but not about long-distance trade. This information can only be obtained from an inter-site comparison on a big enough scale. This study of faunal remains will consider various different scales and approach the data set from different angles. It will do this by zooming in, going from greater distance to more detail.

An INTER-SITE comparison of the entire Iron Age database will investigate differences and similarities in species abundance. Such differences might indicate specific husbandry regimes, which might reflect an adaptation to the environmental conditions in a certain region or be a sign of cultural decisions. These differences might also reflect differing preservation.

The skeletal elements of each species are also analysed in an INTRA-SPECIES comparison. The main purpose of determining the composition of skeletal elements is to examine the profile of each species; this might provide some indication about the processes that led to the composition at the sites. Such an analysis of skeletal element representation can provide useful information about taphonomic or behavioural processes that influenced the composition of the assemblages.

A comparison of the distribution of various elements between species, an INTER-SPECIES comparison, will investigate whether the carcasses of different species are distributed evenly throughout the Iron Age. As with the other two analyses, differences might indicate a different treatment of that species and point towards different husbandry strategies.

It is possible that the results obtained on these different scales of observation appear to contradict each other. In fact Simpson's paradox (see section 8.2) implies that the results obtained on different scales may suggest different relationships or patterns. The inter-species distribution, for instance, might indicate similarities that are not reflected by intra-species studies. This is not caused by a false analysis. The different ways of looking at the data give information on different types of assemblage formation processes. Archaeological data contain information on many different spatial and temporal scales at different levels of aggregation (Olivier 1999). Even though Simpson's paradox makes the interpretation of archaeological data more challenging it also constitutes a great opportunity.

9.5 SPECIES PROPORTIONS

A comparative study of the relative species proportions in the Iron Age assemblage was undertaken, using various graphical methods. The species proportion of an assemblage can be used to make assumptions about the husbandry regimes practised at the time. Differences in the species distribution might suggest environmental conditions or cultural conditions.

Box-and-whisker plots and histograms

Figure 13 shows box-and-whisker plots for the frequencies of each species for both size variables TOTAL NISP (N=35) and SUM NISP (N=41). The plots show that the dataset clearly departs from normality and all species indicate right skewness. Skewness often results in a string of extreme values on one side of the plot; in this case these extremes are within the larger assemblages. Furthermore the median is off-centre with respect to the hinges as well as the box being off-centre in regard to the highest and lowest values. In this case none of the boxes is divided into equal halves but the median is shifted towards the smaller frequencies. While the three species are similar in respect to their location (median) they differ significantly in scale. Small sites with low frequencies clearly dominate the dataset. Pig is represented with the smallest frequencies while sheep has the highest. The extreme values of the TOTAL NISP variables are at least twice as high as for the SUM NISP values, which increases the right skewness in the first set of plots. This is due to the greater number of fragments that could not be assigned to a skeletal element.

The histograms shown in figure 14 confirm the results of the box-and-whisker plots: all three species show a right-skewed distribution and some outliers. The graphs are almost identical for both overall frequency variables and only the TOTAL NISP graphs are given here. These results show that measures of location and spread are meaningless for this dataset, because for skewed distributions the majority of sites lie below the mean, which is thus neither representative nor does it define the location of the distribution (Hartwig & Dearing 1979: 26-27). This was to be expected considering the nature of the dataset.

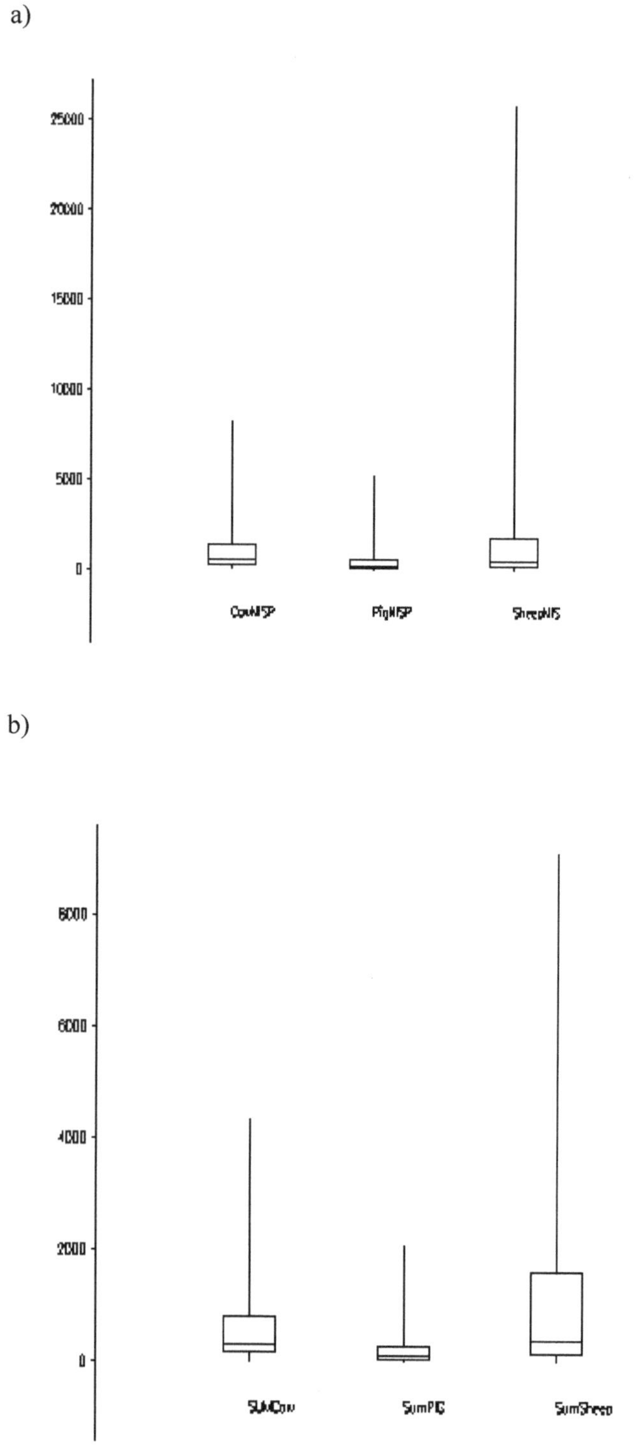

***Figure 13**: a) Box-and-whisker plots for the frequencies of cow, pig and sheep (TOTAL NISP; N=35), b) Box-and-whisker plots for the frequencies of cow, pig and sheep (SUM NISP; N=41).*

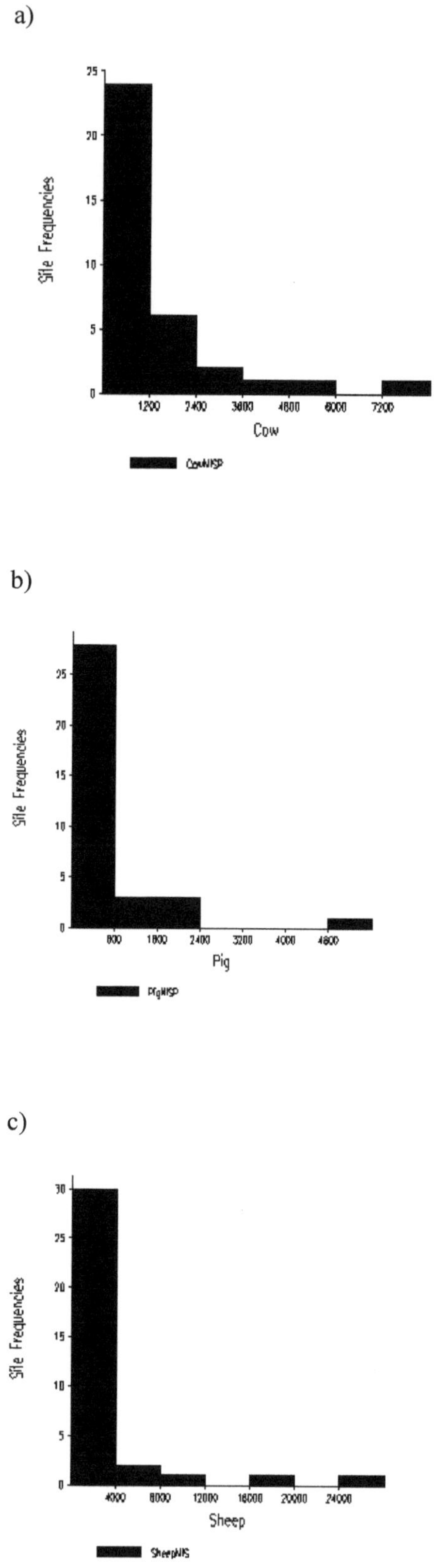

***Figure 14**: Histograms of the TOTAL NISP species frequencies for 35 Iron Age sites a) cow b) pig and c) sheep*

Re-expression of the data

One way of solving the problems created by non-normality is re-expression. This technique provides a means of making the information stored within the data easier to access. It is merely a transformation of the original data into a numeric scale of measurement other than the one in which the original data was recorded. The most common form of re-expression is using logarithms. This transformation alters "[...] the distance between the points along the scale while at the same time maintaining the sequence, or order, of the points" (Hartwig & Dearing 1979: 53).

The distances at the upper end of the scale are decreased while those at the lower end are increased. As a result the intervals between the points are no longer equal, which rules out absolute comparisons, but because the order is maintained a relative comparison is still possible (Tukey 1977: 57-64, Hartwig & Dearing 1979: 52-69). A comparison of the box plots for the frequencies of pig in figure 13a and figure 15 shows how the log re-expression can even out skewed distributions. In the original observations of pig frequencies the samples at the upper end (large assemblages) are too far apart relative to those at the lower end and the shrinkages of the distances within the upper end clearly normalized the data. It should be stressed that re-expression prior to data analysis does not alter the existing relationships between the variables but visualizes them differently, which may help to extract additional information. This technique fits into the concept of openness stressed by exploratory data analysis. If the scale in which the data were observed is taken as arbitrary a re-expression of the original data can be used to discover unanticipated patterns.

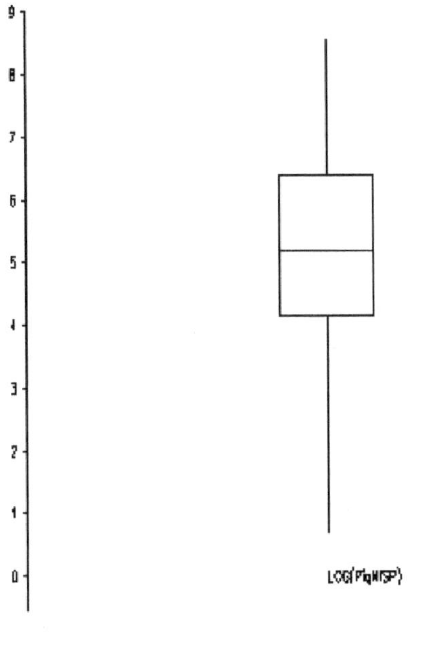

Figure 15: *Box-and-whisker plot for the TOTAL NISP frequencies of pig for 35 British Iron Age Sites on a logarithmic scale*

The species frequencies are transformed onto a $\log_e$ base and again plotted in histograms. The results are shown in figures 16 - 18 and demonstrate the impact of scale very nicely. Compared to the original right-skewed distributions in figure 10.3 the transformed data for all three species seem to fit neatly into a normal distribution when clustered into 6 groups[1] (figure 16a – 18a).

These histograms also demonstrate how Simpson's paradox can influence pattern in the data. When the number of groups on the horizontal axis is increased to 12 all three histograms show multimodal instead of unimodal distributions; the collapsing of categories (creating bigger groups) reduces the diversity in the data and leads to a unimodal pattern. Both give a valid representation of the Iron Age assemblage.

[1] The number of groups is generally calculated as the square root of N (Winder pers. comm.). In this case the number of sites is 46 and the suggested number of groups is 6 or 7.

a)

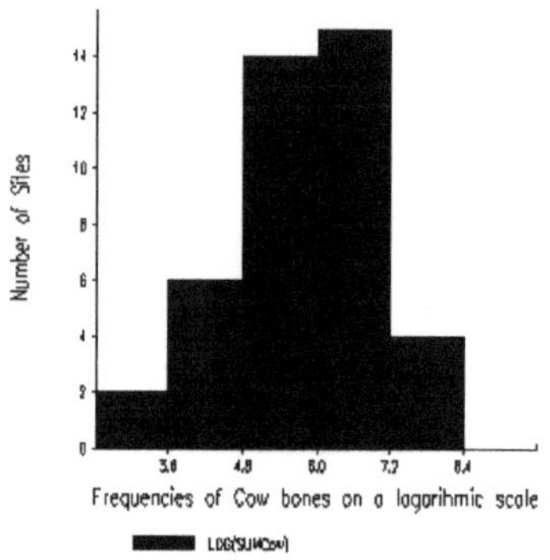

b)

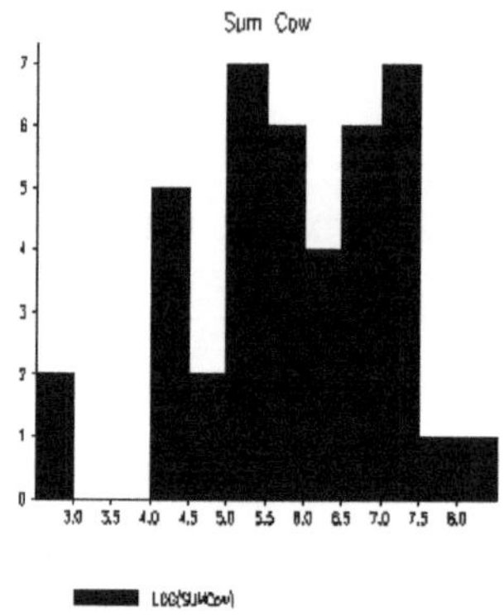

Figure 16: *Histograms for SUM NISP cow frequencies on a logarithmic scale a) data clustered into 6 and b) data clustered in 12 groups*

a)

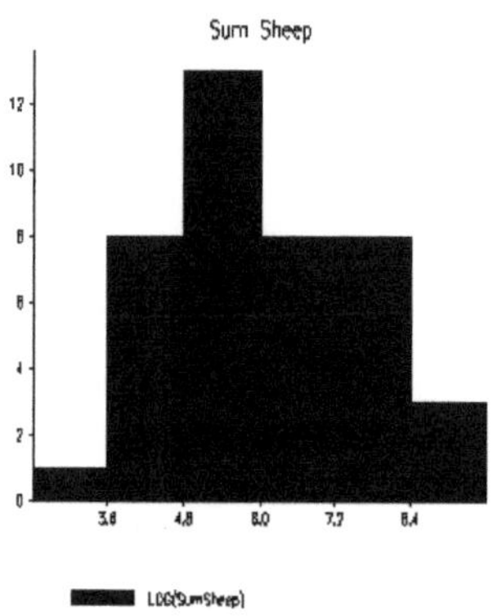

b)

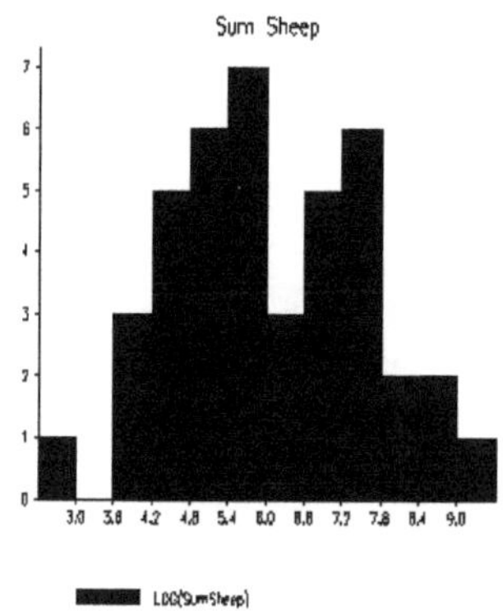

Figure 17: *Histograms for SUM NISP sheep frequencies on a logarithmic scale a) data clustered into 6 and b) data clustered in 12 groups*

a)

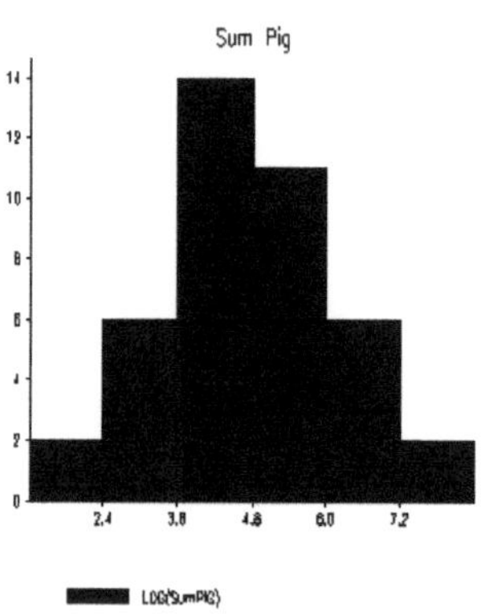

b)

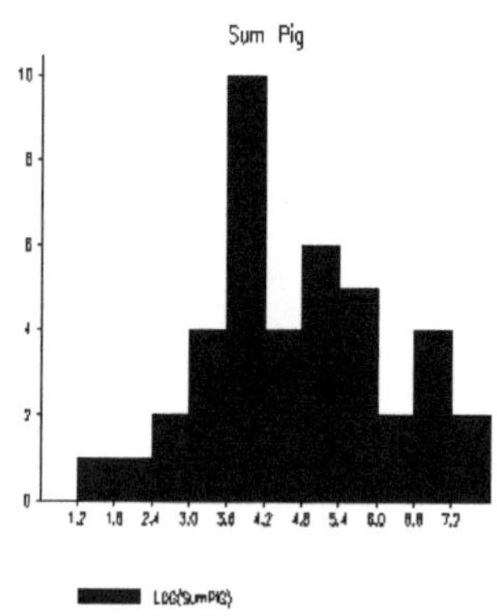

Figure 18: *Histograms for SUM NISP pig frequencies on a logarithmic scale a) data clustered into 6 and b) data clustered into 12 groups*

Scatter plots

The box-and-whisker plots have revealed that pig is less abundant than cow or sheep while the latter has the highest overall frequencies. However individual sites may differ significantly from this overall pattern and scatter plots will be used in order to test whether cow, sheep and pig are represented in the same proportions throughout the entire assemblage. This gives a first impression of the profile of the data and shows the contribution of each species to the overall distribution (figures 10.8 - 10.10). It should be stressed that the points in the scatter plots do not represent Iron Age sites but rather faunal samples. Some of these samples might represent the entire faunal remains from that site but, as the comparison between SUM and TOTAL NISP values has shown, they often represent only a smaller subset. Again the graphs show the comparison for the TOTAL NISP (N=35) as well as for the SUM NISP values (N=41). The patterns suggest a positive linear relationship between the species. The small sites cluster together very tightly while the larger and middle-sized assemblages scatter further and indicate a greater deviation from the general pattern. The scattering is greater when the SUM NISP values are compared. This might be due to the larger sample size of this variable. It also includes the very large site of Owslebury which had to be excluded from the TOTAL NISP counts. The majority of sites contain about twice as many cow bones as pig bones which is why the regression line is tilted to the left. A number of sites contain relatively more pig than cow bones compared with the remainder of the data and three sites appear to have even higher absolute values of pig bones (figure 19).

The comparison of sheep and cow bone frequencies (figure 20) shows that there are generally more sheep than cow bones. However in some of the middle-sized outliers cow bones are more common and sheep are rarer.

Graph 20b appears to show two regression lines, with relatively more sheep bones to the left and sites with about as many cow bones as sheep bones to the right. This is a very interesting pattern because it splits the assemblage into two groups that differ in their species profile. The proportionally higher sheep frequencies on some sites do not necessarily mean that this species contributed more meat to the Iron Age economies than the other domesticates. The meat weight of one cow is significantly higher than that of a sheep. Grant (1991: 458) has suggested an average weight of 30 kilos for an adult sheep and 400 kilos for a cow.

Sheep is also more abundant than pig (figure 21). The regression line is steeper compared to the sheep/cow scatter plot which indicates that the surplus of sheep is even greater in regard to the contribution of pig to the overall assemblage. However some outliers exhibit a deviation from the overall trend by having higher proportions of pig. Because pig bones tend to be less well-preserved than the other two species on these sites, these differences might indicate a greater reliance of the economy on pig products.

The scatter plots show that the species composition revealed by the box-and-whisker plots reflects the composition of the majority of British Iron Age sites where the order of species abundance is sheep, cow and pig. However there are exceptions within the assemblage, especially in the frequencies for cow and sheep, where the rank order is exchanged for various sites. These results are in accordance with the literature on the British Iron Age that assumes a high dependence on sheep, followed by cattle and pig (Cunliffe 2000: 379-404, Hambleton 1999: chapter 6, Maltby 1994). The outliers tend to be middle-sized or large sites. The outlying sites with the higher frequencies for pig bones, for instance, are both large open settlements from the late Iron Age: Ower and Skeleton Green. Because biases caused by quantification methods tend to influence assemblages with small frequencies, they are unlikely to be the cause of these outliers. However sample size might have an impact on the pattern.

Generally the overall pattern for the SUM NISP values suggests two types of sites in regard to the cow/sheep distribution, a pattern that is not visible in the TOTAL NISP plot.

a)

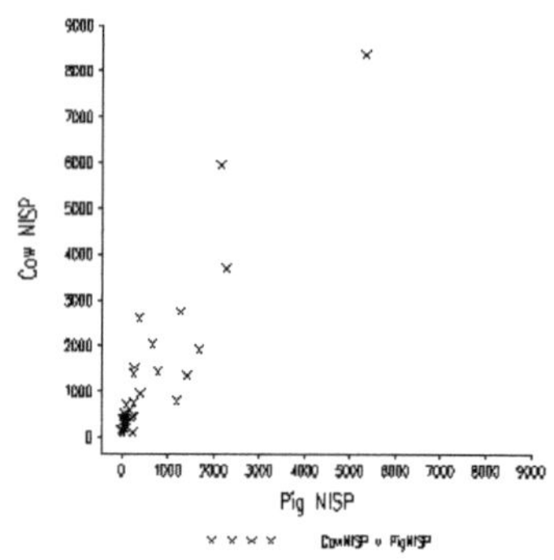

b)

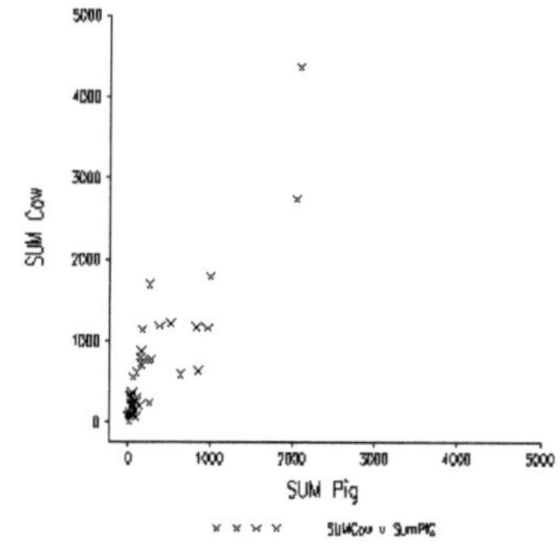

Figure 19: Relationship between the frequencies of cow and pig bones; a) Correlation between TOTAL NISP (N=35) and b) SUM NISP frequencies (N=41)

a)

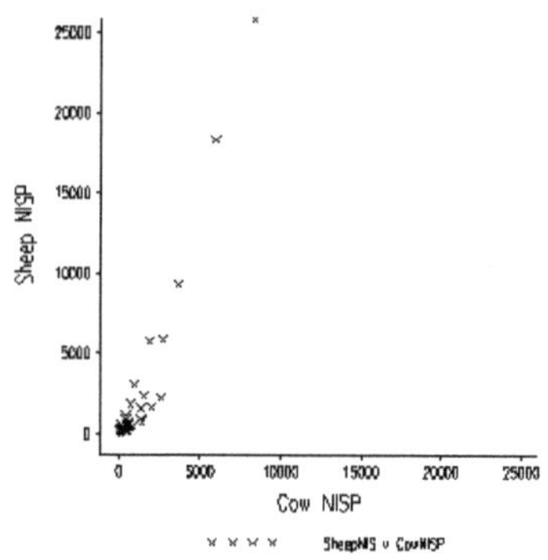

b)

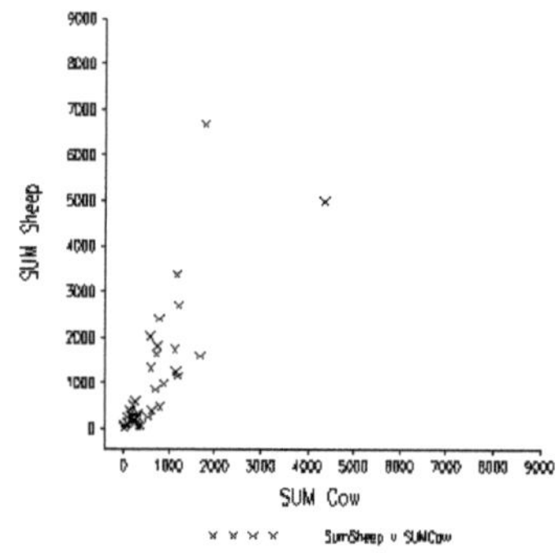

Figure 20: Relationship between the frequencies of cow and sheep bones; a) Correlation between TOTAL NISP (N=35) and b) SUM NISP frequencies (N=41)

a)

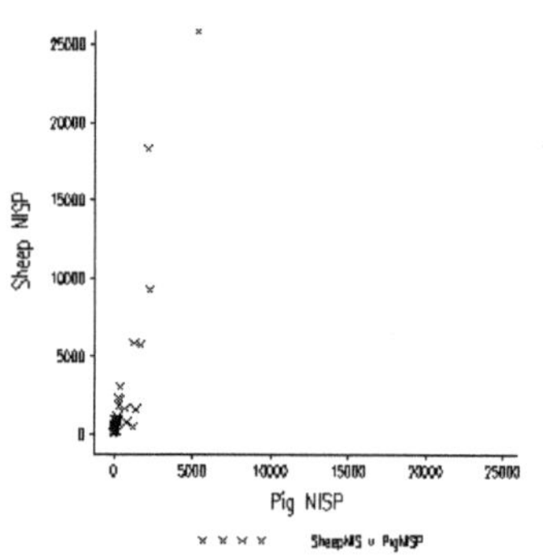

b)

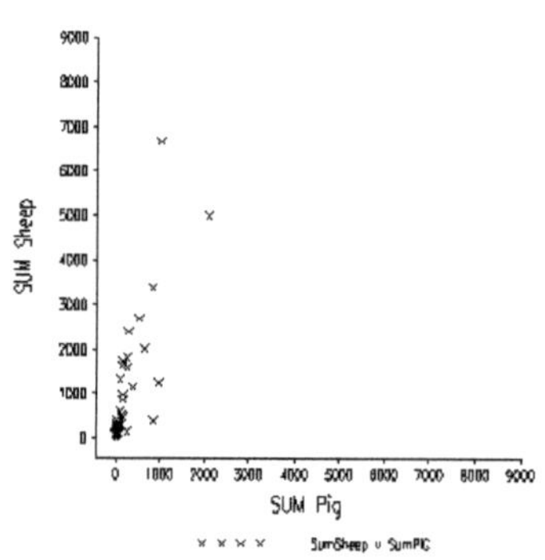

Figure 21: Relationship between the frequencies of pig and sheep bones; a) Correlation between TOTAL NISP (N=35) and b) SUM NISP frequencies (N=41)

9.6 RANKING OF SKELETAL ELEMENTS: FOR THE OVERALL ASSEMBLAGE

Having compared the frequencies for each species the graphs are then used to test for variation within the composition of skeletal elements of each species. Such a comparative study can provide useful information both about the processes influencing the composition of the assemblages as well as the general level of preservation. As discussed before, it can be assumed that the distribution of skeletal elements found at a site differs from complete skeletons for various reasons (section 8.2). Depending on their history and constitution, certain skeletal elements are expected to appear at higher frequencies than others. Ideally the state of fragmentation would be determined for each site and sites where taphonomic processes are likely to be responsible for the observable skeletal element distribution would be identified. However MNI data is only available for a very few sites and bone weight is not given at all. Hambleton (1999: chapter 4) has argued that NISP frequencies should not be used to compare the overall level of preservation of different assemblages, because differences in fragmentation between the sites which cannot be detected with this quantification method may obscure the results. She could demonstrate that in some cases the MNI and NISP frequencies of the same site showed different patterns for the overall abundance of skeletal elements. However, she also points out that both methods showed a similar pattern for some sites and that NISP data "provides an indication of which elements were and were not counted in the initial recording of the material" (1999: 27). The lack of better data does not give a choice. The information from the NISP frequencies in regard to skeletal element representation is therefore not rejected but analysed in order to recognise trends in the Iron Age assemblage. However the interpretation of the results will consider the difficulties and biases that may result from the use of NISP frequencies.

Figures 22 - 24 show the frequencies of each recorded skeletal element for cow, pig and sheep for 46 British Iron Age sites.

The mandible is clearly the most abundant element for all three species. The teeth are very robust and easily identifiable even when the jaw is fragmented and can therefore be expected to appear in high frequencies (Brain 1981: chapter 2, Payne 1972b). The difference between the mandible and other skeletal elements is highest within the pig assemblages. The post-cranial skeleton of pig is known to have lower survival rates than the skeletons of the other main domesticates. This is due to the slaughter of relatively young animals which decreases the chances of bone survival (Hambleton 1999: chapter 4). This pattern could thus be explained through taphonomy.

The next most abundant elements for **cow** (figure 21) are the scapula, tibia and humerus. All three bones score relatively high in Binford's modified utility index (1978: 74) and have high meat contents. The frequencies of vertebrae, pelvis and radius appear next in the rank order. Ribs have a low frequency which is in accordance with their fragile nature, while the low frequencies for the phalanges might represent recovery biases (see section 9.2.3).

While there are generally fewer **pig** bones (figure 23), the rank order of frequencies is very similar to the cow bones. The second most abundant elements after the mandible are also scapula, humerus and tibia, followed by pelvis and vertebrae. For pig the ribs are the least abundant element.

The distribution of **sheep** bones (figure 24) differs from the other two species: the tibia rather than the scapula is the second most abundant element, followed by the humerus and vertebrae. The scapula appears in significantly lower frequencies within the sheep assemblages than within the pig and cow bones but the tibia has higher frequencies.

Generally the distributions of pig and cow bones show greater similarities while sheep frequencies differ. It could be argued that cows were used for purposes other than food, for instance traction or ritual, while sheep husbandry concentrated purely on subsistence. However this point might explain the differences in the distribution of cow and sheep bones but it would not explain the similarities between cow and pig.

9.7 RELATIONSHIP BETWEEN INDIVIDUAL BONES AND THEIR OVERALL SPECIES DISTRIBUTION

In order to get a better understanding of the data set, every skeletal element is plotted against both species frequencies. This shows how each bone relates to the overall species distribution and if there are any differences. A strong correlation between the variables suggests that the relative proportions of these particular bones are the same throughout all Iron Age sites. In that case there are no real differences between the sites. A weaker correlation on the other hand might indicate patterns caused by taphonomy or human behaviour and is therefore of great interest for this study.

Each element was plotted against both overall frequencies for its species (see section 9.3). Due to limited space not all of them are shown here. For **cow** the choice of overall species representation shows an impact on the results: the relationship between most elements and the SUM NISP values resembles a straight line with a weak correlation (examples are given in figures 25b & 26b) while most of the elements (femur, metacarpal, metatarsal, humerus, pelvis, scapula and tibia) appear to show two linear relationships when plotted against the TOTAL NISP values (figures 25a & 26a).

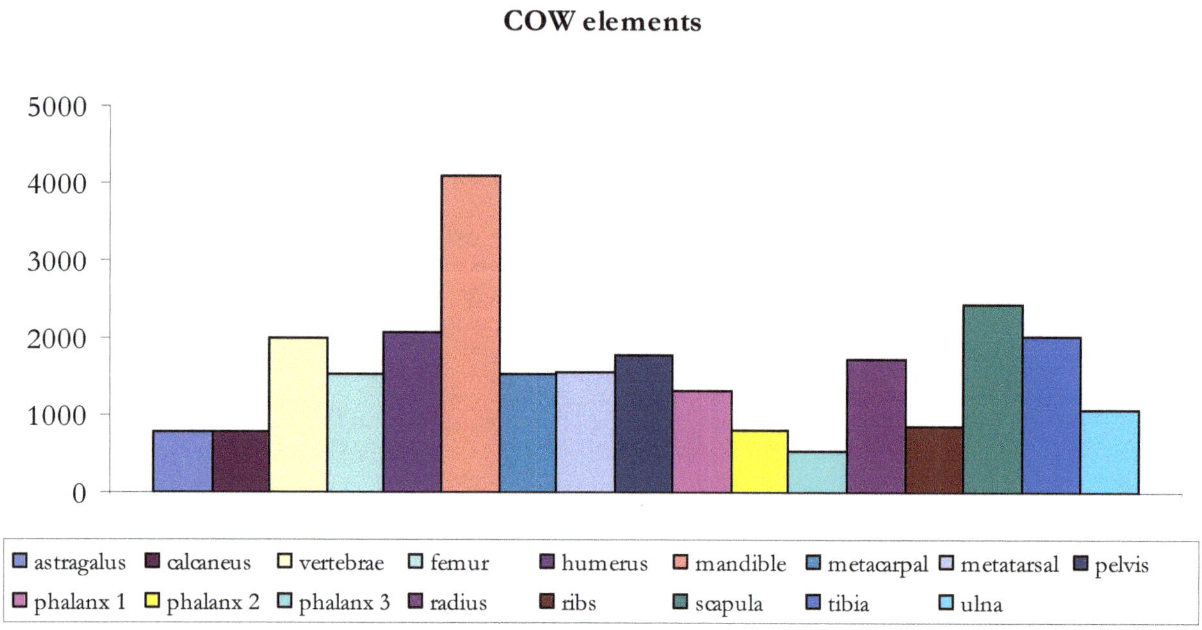

Figure 22: *Frequencies of the skeletal elements for cow across all sites*

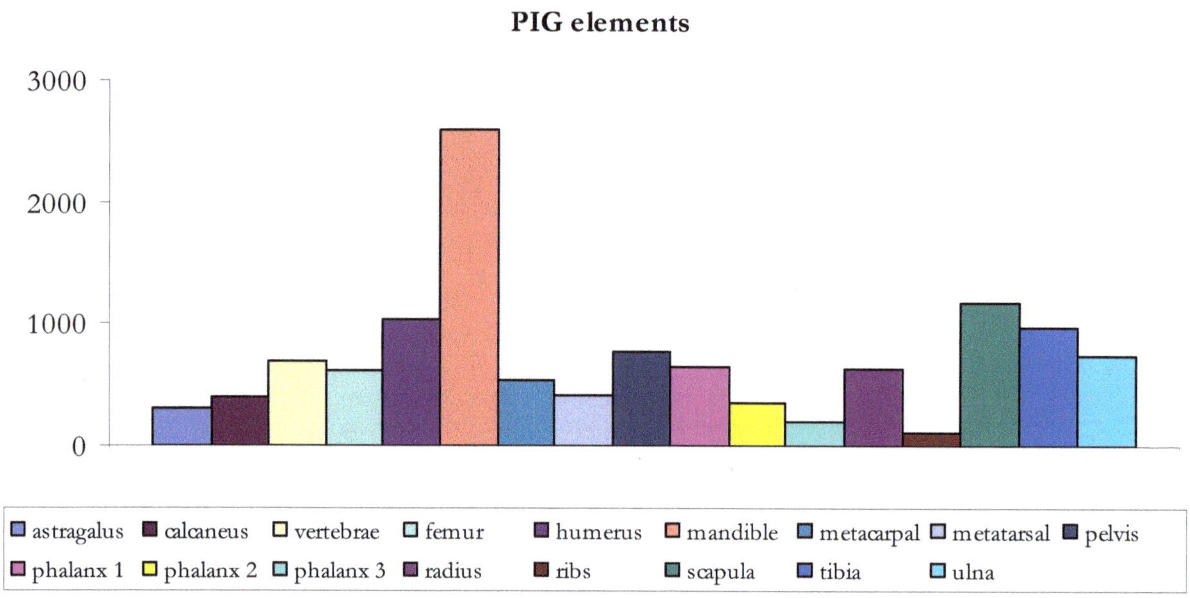

Figure 23: *Frequencies of the skeletal elements for pig across all sites*

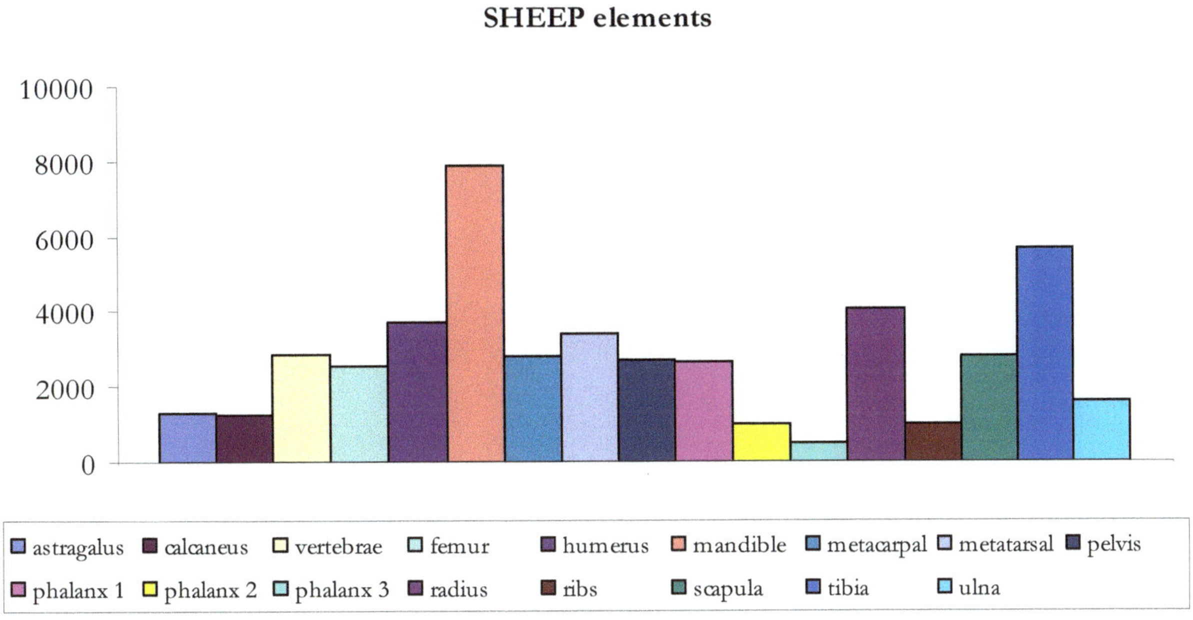

Figure 24: *Frequencies of the skeletal elements for sheep across all sites*

Thus the TOTAL NISP values appear to be more sensitive in identifying patterns than the SUM NISP values. This is surprising because in the species comparison the SUM NISP values showed greater scattering than the TOTAL NISP values (see section 9.5). But while the SUM NISP variable includes more sites, the TOTAL NISP frequencies contain higher and more extreme values. Since the two regression lines are easier to distinguish for larger assemblages the inclusion of extremely high frequencies might be responsible for the pattern in figures 25a and 26a.

The **sheep** elements tend to show the same pattern independently from the choice of species frequencies. The majority of sheep elements plot in a straight line, indicating that the distribution of sheep bones is the same throughout all the Iron Age assemblages (figure 27). However some of the bones (vertebrae, tibia, metatarsal, the phalanges 1 to 3 and radius) show two linear relationships. Figure 10.16b indicates that there might be two different types of sites within the assemblages, some with proportionally more sheep tibias and others with proportionally less.

For **pig** most elements exhibit a positive linear relationship to the overall species distribution but often with a weak correlation for the larger assemblages. This might indicate the existence of two regression lines, but the pattern is not as obvious as in the other two species (figure 28b). Similar to the cow plots the correlation is generally stronger for the SUM NISP values while the TOTAL NISP values show more of a pattern (figure 28a).

These scatter plots have shown that the skeletal elements are not distributed evenly throughout the entire Iron Age assemblage. Some of the cow and sheep elements exhibit two regression lines, which indicates that there are two types of assemblage which differ in their proportion of these elements. This result is very interesting because it could represent taphonomic processes or differential human behaviour. The pattern is less common for the pig elements but still visible in some of the plots.

a) b)

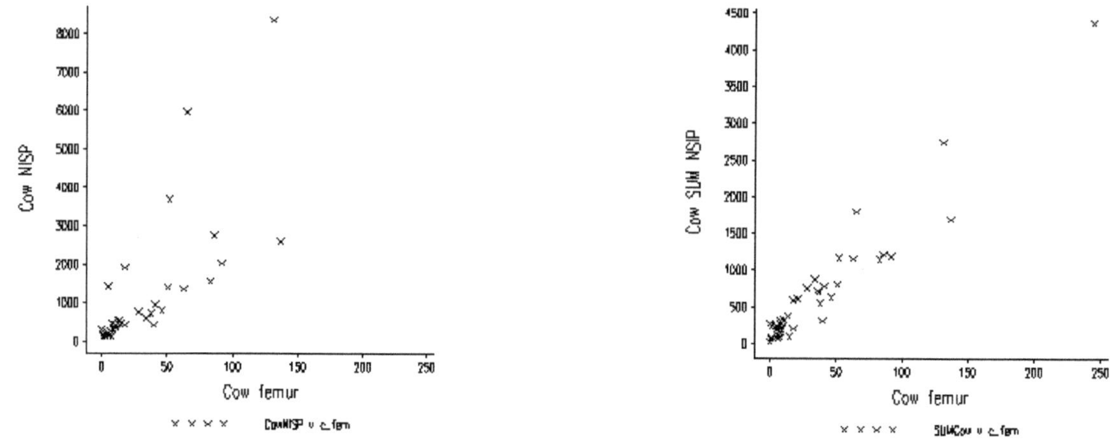

Figure 25: Relationship between cow femur and both overall species frequencies a) TOTAL NISP and b) SUM NISP. The TOTAL NISP plot appears to have two different linear relationships.

a) b)

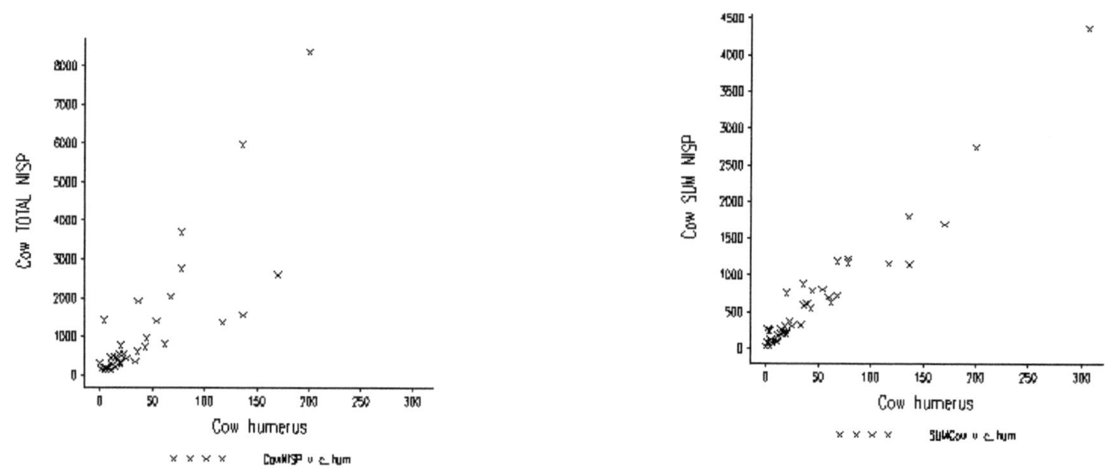

Figure 26: Relationship between cow humerus and both overall species frequencies a) TOTAL NISP and b) SUM NISP. While graph b shows a strong correlation, graph a shows two different linear relationships.

a) b)

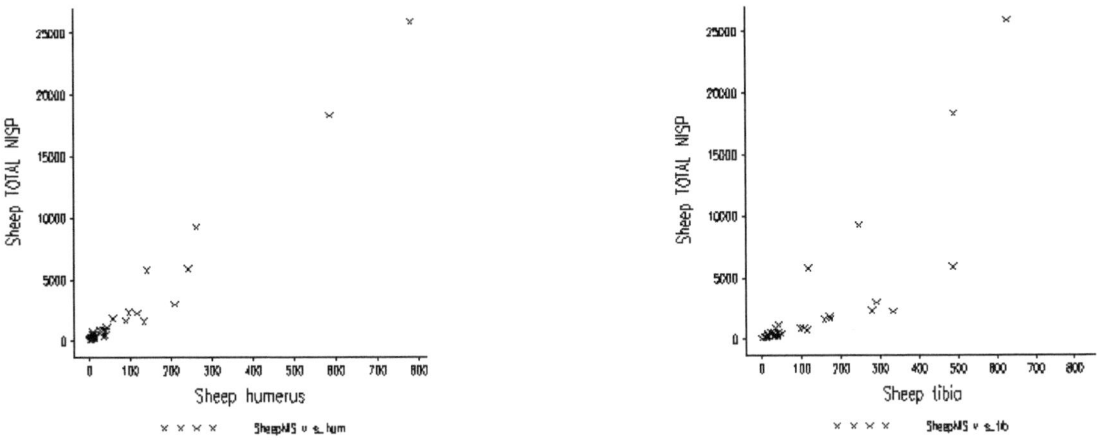

Figure 27: relationship between the TOTAL NISP sheep values and a) sheep humerus and b) sheep tibia. While graph a) lacks a pattern graph b) shows two linear relationships.

a) b)

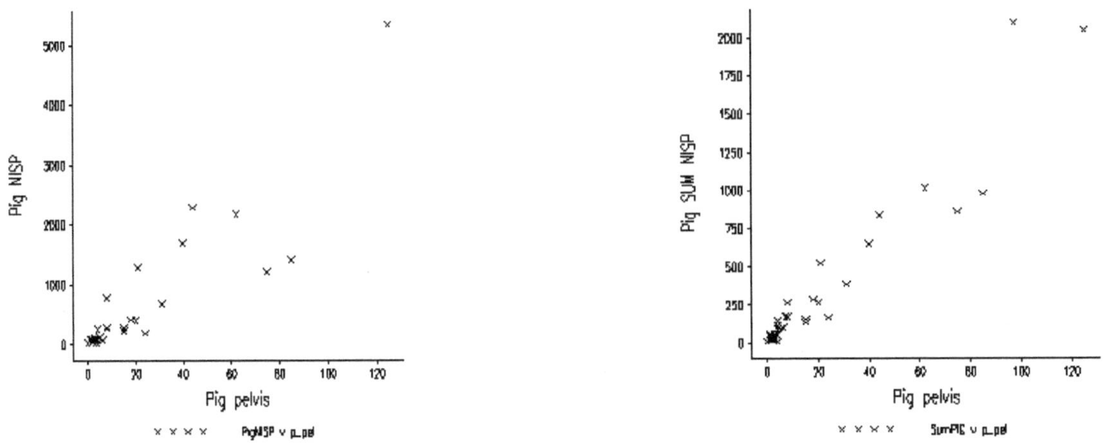

Figure 28: Relationship between the pig pelvis and a) the pig TOTAL NISP values and b) the pig SUM NISP values

9.8 RELATIONSHIP BETWEEN SKELETAL ELEMENTS OF ONE SPECIES

Next, the frequency of one skeletal element is compared to another element of the same species. This type of plot helps to examine the profile of each species; again a close correlation between two elements suggests that their relative proportion is the same across the entire assemblage. A weak correlation suggests differences between individual sites and could point towards taphonomic or behavioural processes that influenced the composition of the assemblages.

a)

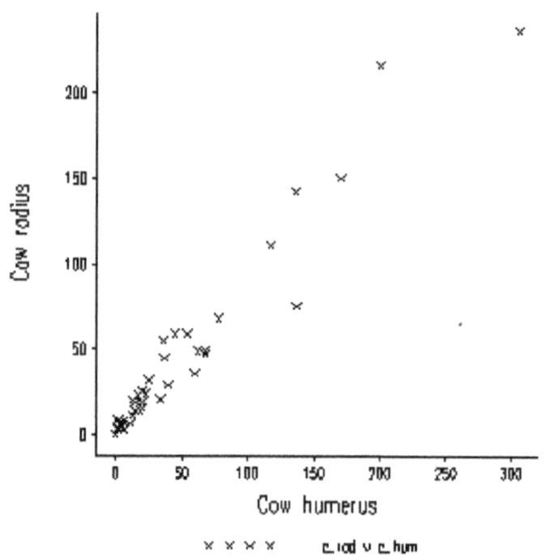

b)

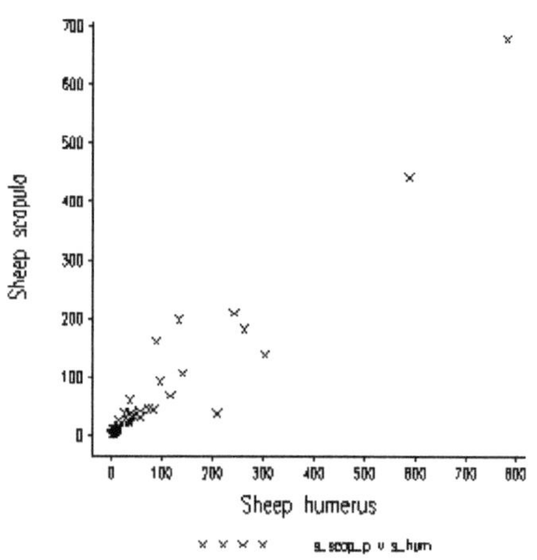

Figure 29: Scatter plots of skeletal elements that cluster together according to Binford's study of Inuit butchery practices (1981: 93). a) Scatter plot of cow radius and humerus and b) scatter plot of sheep scapula and humerus.

First, elements that can be expected to enter the archaeological record together, according to Binford's study of Inuit butchery practices (1981: 93), are plotted against each other. The low number of wild animals found in faunal assemblages suggests that hunting did not play an important role in the subsistence strategies of the British Iron Age. It is thus unlikely that the faunal remains of this period fit into the pattern observed for societies that rely mostly on hunting. However the close correlation of skeletal elements might give us an indication about the slaughtering practices at the site. The graphs of limb elements show a strong correlation between the bones, which indicates that their relationship is the same throughout all sites (figure 29).

In the next step the skeletal elements are paired randomly. Figure 30 shows the scatter plots for pig mandibles and the first phalanges. Both elements can be expected to enter the archaeological record separately. The majority of sites seem to fit into a perfect regression line, but there are four outliers with relatively higher frequencies of phalanges. This could possibly show two regression lines, a pattern that has been observed before when individual elements were plotted against the overall species frequencies. Some sites have more pig mandibles while others have higher frequencies for the first pig phalange.

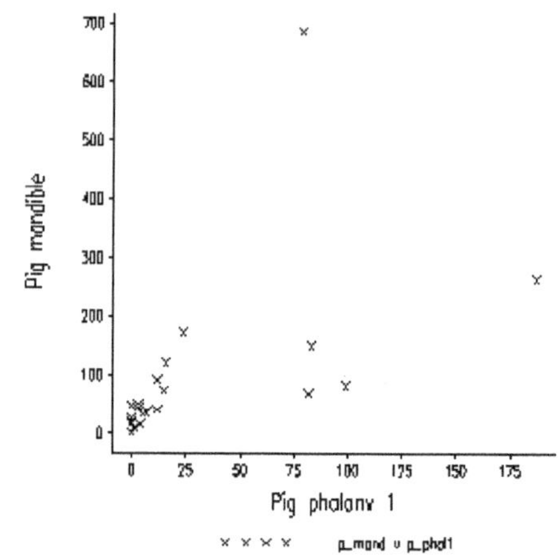

Figure 30: Scatter plots for pig mandible and first phalanges

9.9 UNIVARIATE REPRESENTATION OF SKELETAL ELEMENTS

While scatter plots give a bivariate representation of the data, histograms show a univariate picture. Because the patterns in the histograms for the species frequencies were dependent on the number of categories (see section 9.5), the axis is also varied for the skeletal histogram representation. Again the distribution of the data differs according to the resolution of the data (see figure 31 & 32). This is related to Simpson's paradox. The examples show the distribution of both the pig mandible and the sheep tibia, in each case the data is grouped in 7 as well as 12 classes. For the pig mandible the graphs show a unimodal distribution, while the higher resolution shows a multimodal pattern. It is not possible to decide which pattern describes the distribution of pig mandible better as both graphs give an equally valid representation of the data.

a)

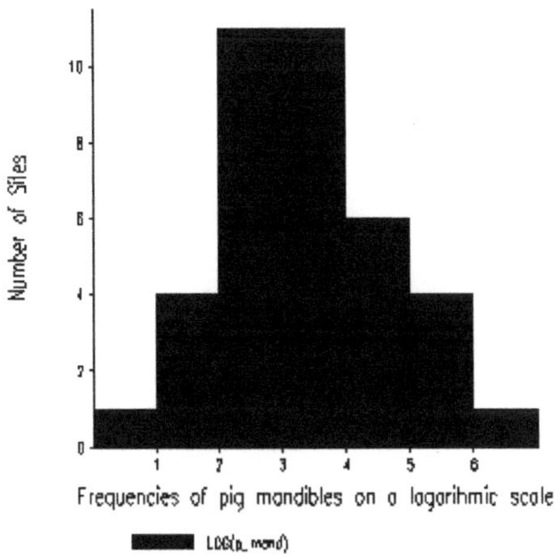

b)

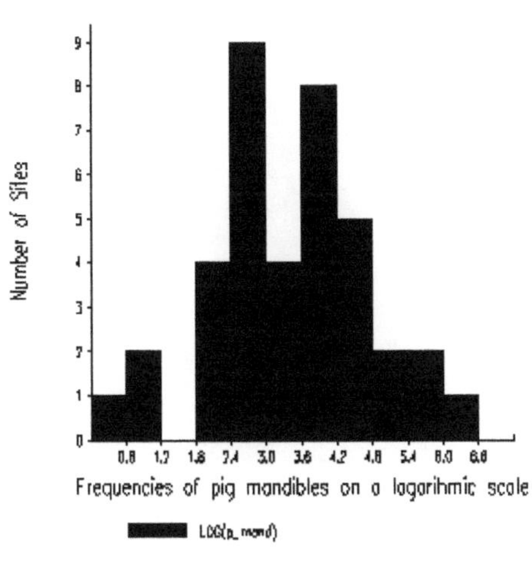

__Figure 31__: Histogram for pig mandible on a logarithmic scale a) for 7 groups and b) for 12 groups

a)

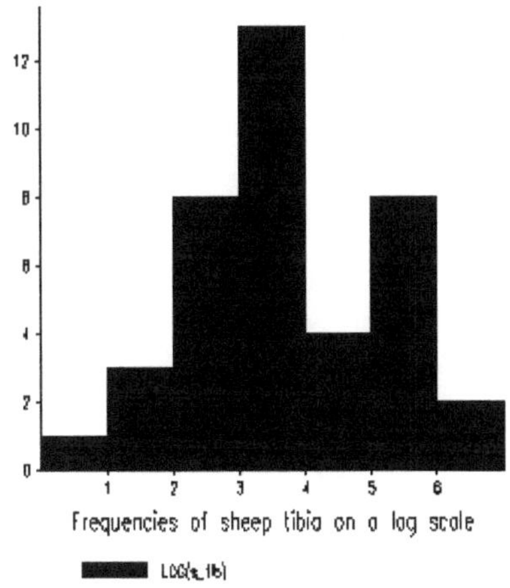

b)

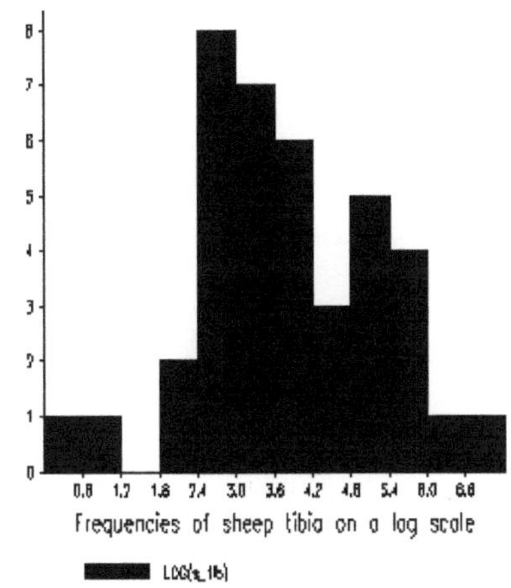

__Figure 32__: Histogram for sheep tibia on a logarithmic scale a) for 7 groups and b) for 12 groups

9.10 RELATIONSHIP BETWEEN SKELETAL ELEMENTS OF DIFFERENT SPECIES

In the last section of this exploratory data analysis the skeletal elements of different species are compared with each other. This is done in order to test whether the carcasses of different species are distributed evenly throughout the Iron Age. We already know that the most frequent skeletal elements of pig and cow appear in the same rank order, but is this overall pattern valid for each individual site? And how do the rank order differences for sheep contribute to the relationship between the species?

Figure 33 shows the relationship between the second most abundant element for cow, the scapula, and the pig (a) and sheep tibia (b). The frequencies for tibia rank differently within sheep and pig; it is more abundant for sheep than for pig which shows in the scatter plots which tilt in opposite directions. Both scatter plots show a regression line with a weaker correlation for the middle-sized and large sites. Thus the differences in rank order of the sheep and pig tibia appear to be evenly distributed throughout the assemblage.

a)

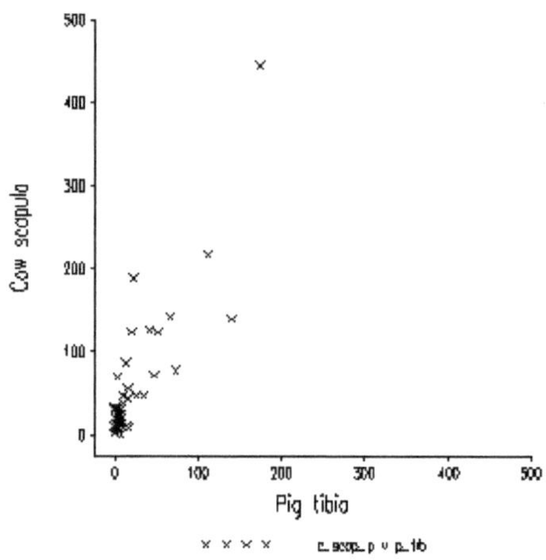

b)

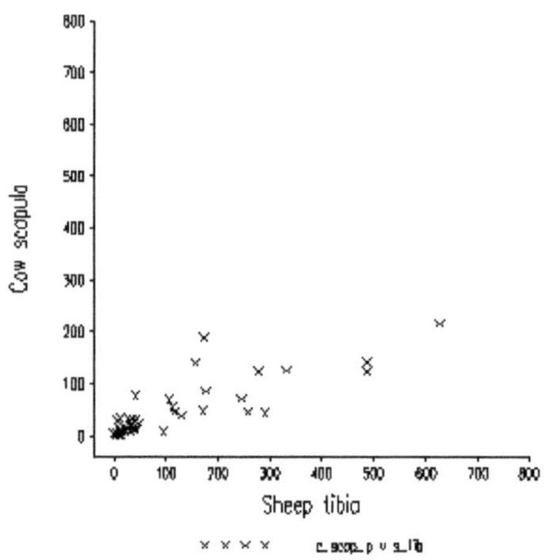

Figure 33: *Relationship between cow scapula and a) pig and b) sheep tibia*

The next set of scatter plots determines the relationship between cow scapulae and pig humeri or sheep humeri respectively (figure 34). Both plots could show two regression lines but the pattern is clearer for the sheep bones. It might suggest that there are two types of assemblages, one with proportionally more sheep humeri and one with proportionally more cow scapulae. The same pattern appears when the sheep scapula is plotted against the cow humerus (figure 35).

a)

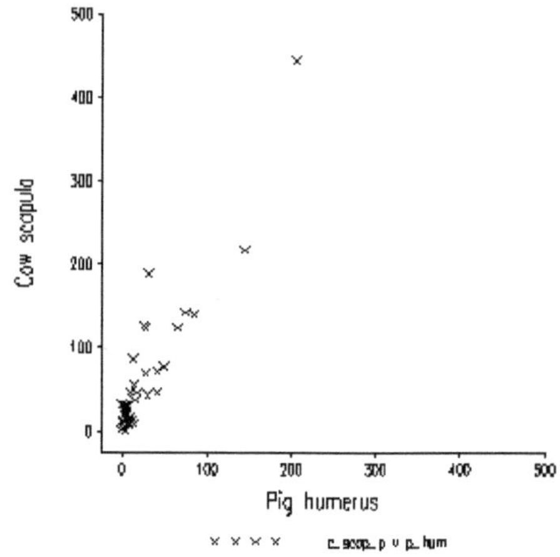

b)

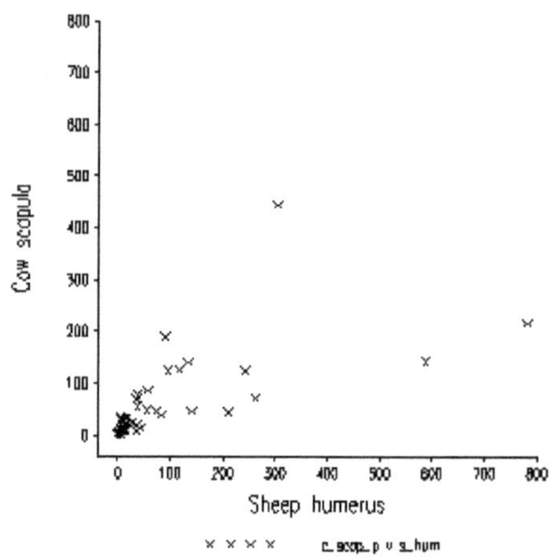

Figure 34: *Relationship between cow scapula and a) pig and b) sheep humerus*

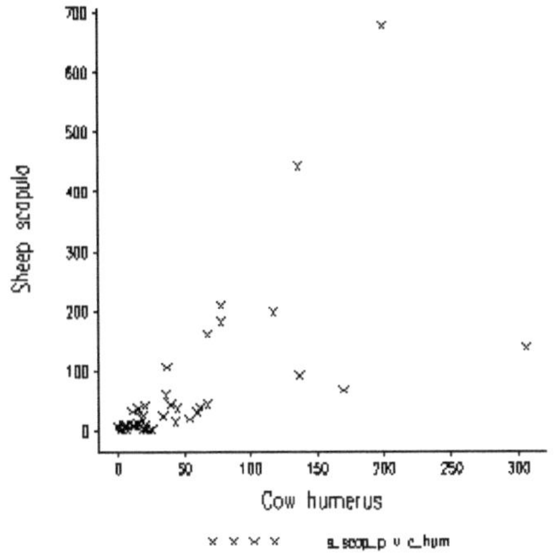

***Figure 35**: Relationship between sheep scapula and cow humerus*

Figure 36 shows the relationship between the pig mandibles and a) cow vertebrae and b) sheep vertebrae. Both graphs show a weak overall correlation with an internal pattern. This suggests that there might even be three types of assemblages.

9.11 CONCLUSIONS

It has been demonstrated that a purely graphical data analysis can provide useful information about inter- and intra-site specific patterns in animal husbandry regimes. The dataset is dominated by assemblages with small frequencies. The inter-site comparison has shown that sheep was the most abundant species in the majority of assemblages, followed by cow and pig. However there are exceptions to this overall pattern and cow is sometimes the most abundant species.

The analyses of the skeletal element rank order have shown that the distribution of sheep differs from the other two species, while cow and pig show great similarities. This might indicate that the sheep bone assemblage formation was different and thus led to differences in the recovery of various skeletal elements. The scatter plots have indicated that there is a positive linear relationship both between each skeletal element and the overall species distribution as well as between two elements of one species. One pattern that kept reappearing throughout various analytical perspectives is a low overall correlation with an internal structure suggesting two types of assemblages. This pattern is easier to observe for the large assemblages. This might be due to the nature of the pattern – it is more difficult to separate two regression lines when they are very close to each other – or suggest a pattern within the largest assemblages. However these differences cannot be explained through sample size alone, because both regression lines contain large assemblages. The pattern might be a matter of *profile* and its exact nature could be of great interest for the analysis of changes in behaviour.

Based on these results statistical methods will be used to gain a more detailed insight into the nature of these outliers. A Principle Component Analysis will be used to investigate whether there are recognisably different groups within the dataset and what it is that differentiates them, human behaviour or taphonomic processes.

a)

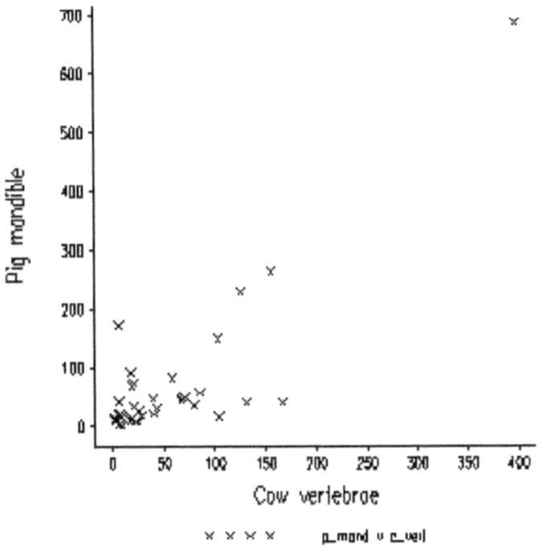

b)

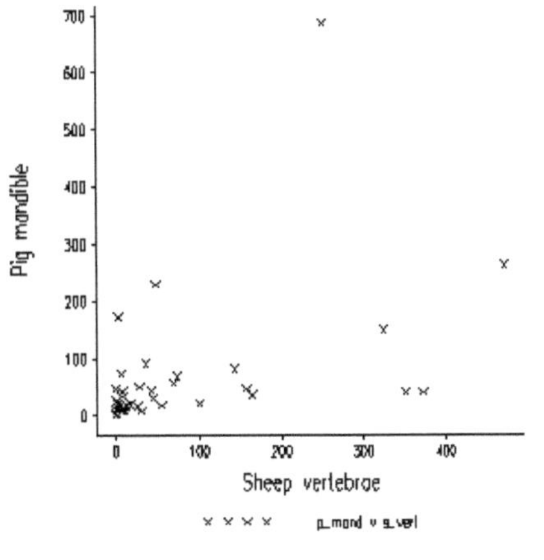

***Figure 36:** Relationship between pig mandibles and a) cow and b) sheep vertebrae*

10. INTERPRETATIVE DATA ANALYSIS: PART II

This chapter will further investigate the patterns detected by the exploratory data analysis and aims to identify possible causes for the variation using Principal Component Analysis as well as other techniques.

10.1 PRINCIPLE COMPONENT ANYLSIS

The exploratory data analysis has shown that the datasets lack normality, which is not surprising considering the number of very small samples within the dataset as well as the size differences between both extremes. The numbers of identified skeletal elements range from 35 to 13917 over the 46 Iron Age sites. Various visual inter- and intra-site comparisons have suggested that there are differences between the British Iron Age sites. However sample size can not account for all the observable differences, because of the inter- and intra-site variation between the large assemblages.

Based on these results a Principle Component analysis (PCA) is used to clarify the nature of the observed patterns. PCA is a multivariate method used to reduce large datasets of interrelated variables or to detect structure in the relationship between these variables. The method reduces the dataset to a number of vectors or principal components (PC) with minimal distortion of the original information. These vectors are placed into a q dimensional space, where q equals the number of variables included in the analysis. The vectors are independent from each other and the variance of all elements is maximized (Schulze 1990: chapter 4.3, Jolliffe 2002: chapter 1, Toit et al 1986: 114-121). Theoretically there may be as many principal components within a PCA as there are variables. However, since PCA is used to reduce the number of variables the number of vectors should be smaller and, usually, the first few principal components jointly explain most of the overall variation while the remaining factors only make a slight contribution. There are no objective criteria for choosing the number of principal components to represent the data. A common method is to select the cumulative percentage of variation that should be described by the PCs, usually 80 to 90%. The smallest number of PCs that is needed to explain this variance is the number of PCs that are chosen to represent the dataset.

While this analysis would ideally include all the available information, PCA is easily dominated by missing values and, due to the high numbers of missing values in the original faunal data set, its structure had to be altered in order to increase the number of sites that could be used. This procedure is a trade-off between keeping the reduction of the original data to a minimum and maintaining a large comparable database. Three new variables were created: the values for the variables phalange 1, phalange 2, phalange 3 and phalanx are summarised into one variable (x_phal) as phalanx. Metatarsal and metacarpal are combined into a so-called x_foot variable and radius and ulna frequencies are also joined into one variable (x_radul). For some of the sites this accumulation had already been done in the original report. The accumulation of frequencies from more than one skeletal element into one variable creates higher NISP values which should be considered during the data interpretation. But, again, this is justified through the necessary trade-off between usefulness and completeness. Ribs, vertebrae, astragalus and calcaneum were excluded from the analysis due to their high numbers of missing values. For Heathrow LIA-RB, Eldon Seat and Hartigans no NISP values are given for the skeletal elements of pig and for Baldock LIA there are no data for sheep. These sites had to be excluded from the analysis (see appendix 2).

There are now 9 skeletal elements for each species which can be analysed statistically, some of which summarize the frequencies of more than one variable (see table 4). Out of the 46 British Iron Age sites in the original database 37 could be included in the Principal Component Analysis. They are listed in table 5. A PCA was calculated based on a variance-covariance matrix and the results are given in table 3. The first two principle components describe 94.77% of the overall variation.

Principal Component	1	2	3	4
Percentage variation explained	83.07	11.70	2.06	1.20

***Table 3**: PCA for 37 British Iron Age sites with 9 skeletal elements for each species*

10.2 GRAPHICAL REPRESENTATION OF THE PCA RESULTS

A multivariate dataset can be represented through the loadings of the first two principal components. Such a scatter plot gives the best possible two-dimensional representation of the data and can be very useful in detecting patterns. "If a good representation of the data exists in a small number of dimensions then PCA will find it, since the first q PCs give the 'best-fitting' q-dimensional subspace" (Jolliffe 2002: 78). Since the components are uncorrelated a linear relationship between the PCs is impossible (Jolliffe 2002: chapter 5). Such a PC plot can also help to identify multivariate outliers or "observations that are a long way from, or in inconsistence with, the remainder of the data" (Jolliffe 2002: 232). Such outliers might be difficult to detect in datasets with many variables, as observations that are within the expected distribution for each of the original variables can still be extreme because of an unusual *combination* of values which contradicts the correlation structure of the remainder of the data. For instance if the height and weight of school children is measured, 175 cm is a normal height for an older child and 25 kilos is not an unusual weight for a younger one, however the combination of both observations within one individual would be atypical (Jolliffe 2002: chapter 10).

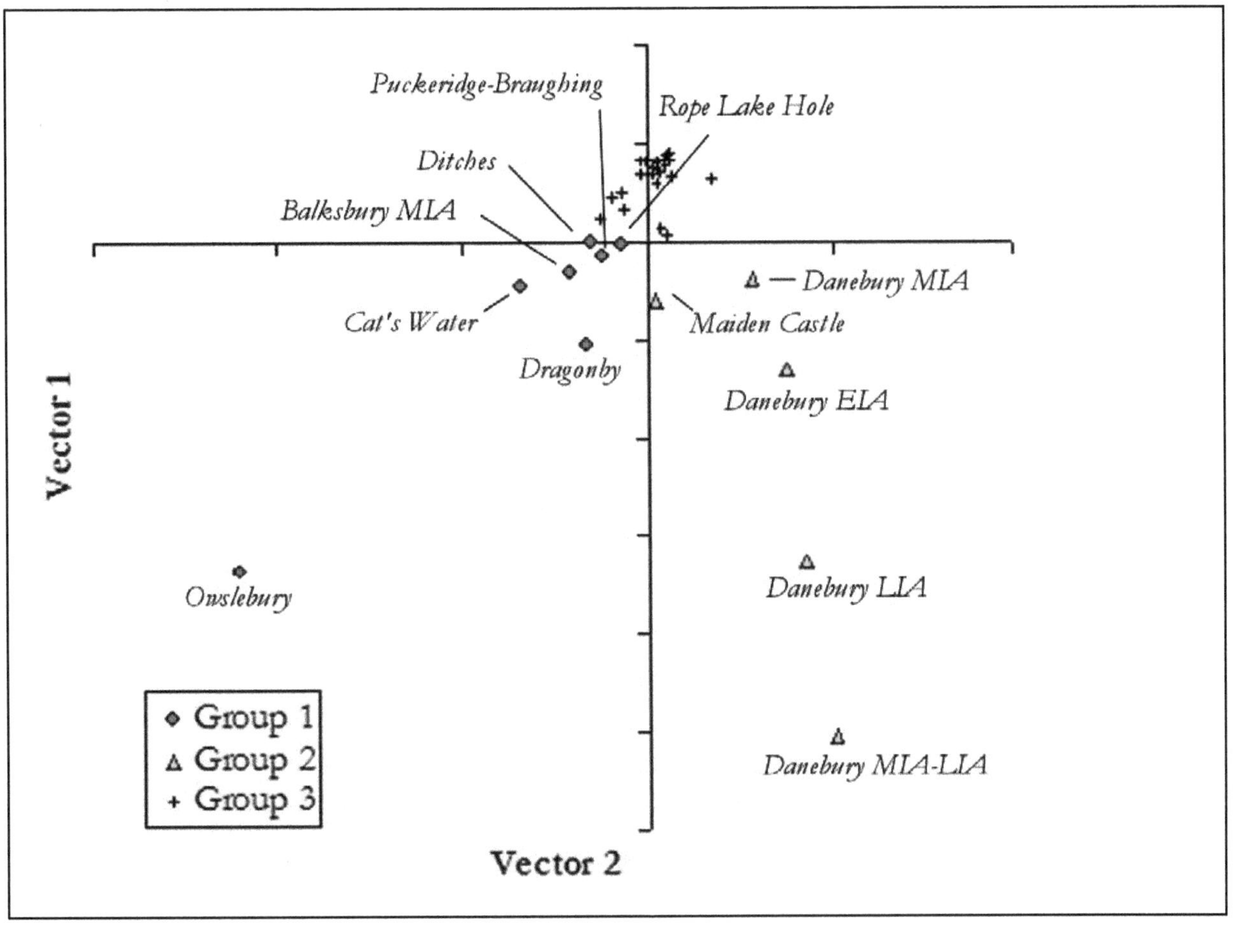

Figure 37: *Scatter plot for the vector loadings of the first two components for the Iron Age sites*

Site representation

Within such a two-dimensional representation, the vectors found by the PCA constitute the x and y axes. The component loadings give the correlation coefficient between each variable and the principal component or, in other words, the distance from each datum point to the centre of the coordinate system. PC loadings are obtained both for all the row variables (in this case British Iron Age sites) as well as for the column variables (skeletal elements). Figure 37 shows the *site* representation for a two factor solution.

The majority of sites cluster tightly together but there are clear outliers, one group clusters on the left arm of the distribution and 5 sites form a line on the right towards the bottom of the graph. One site, Owslebury, is rather isolated at the extreme left of the plot. The graph shows a so-called 'horseshoe' distribution. This shape seems to confirm the pattern formed by the two regression lines visible in some of the scatter plots in chapter 9. To test this assumption the information of both the graphical and the statistical analysis is combined by marking the outlier sites on the scatter plot comparing the cow and sheep frequencies (see figure 38).

The plot shows that the outliers identified by both methods are identical and that the same sites cluster together in one regression line in both distributions. Thus the pattern suggesting two types of assemblages is confirmed by the statistical analysis of the data. This particular shape of two regression lines will be referred to as 'bifurcation' from now on. The positions of the sites outside the main cluster (figure 37) indicate that they are different from the rest of the assemblages and the history behind this outlier pattern as well as the exact nature of these differences might provide us with insight into the behaviours of Iron Age people.

The graphical analyses suggested that most outliers are very large assemblages. Table 5 compares the SUM NISP values of all sites and verifies that the outliers have the highest bone frequencies within the overall assemblage. This first result might point towards the product of earlier innovations because large sites have often been associated with complexity, incipient urbanisation and progression (Hill 1995, Sherrat 1996), factors that are supposed have eased Romanization.

Although sample size is known to have a distorting effect on datasets and it is possible that the outliers simply represent quantification biases, the vector representation suggests that this is not the only cause of the observed pattern.

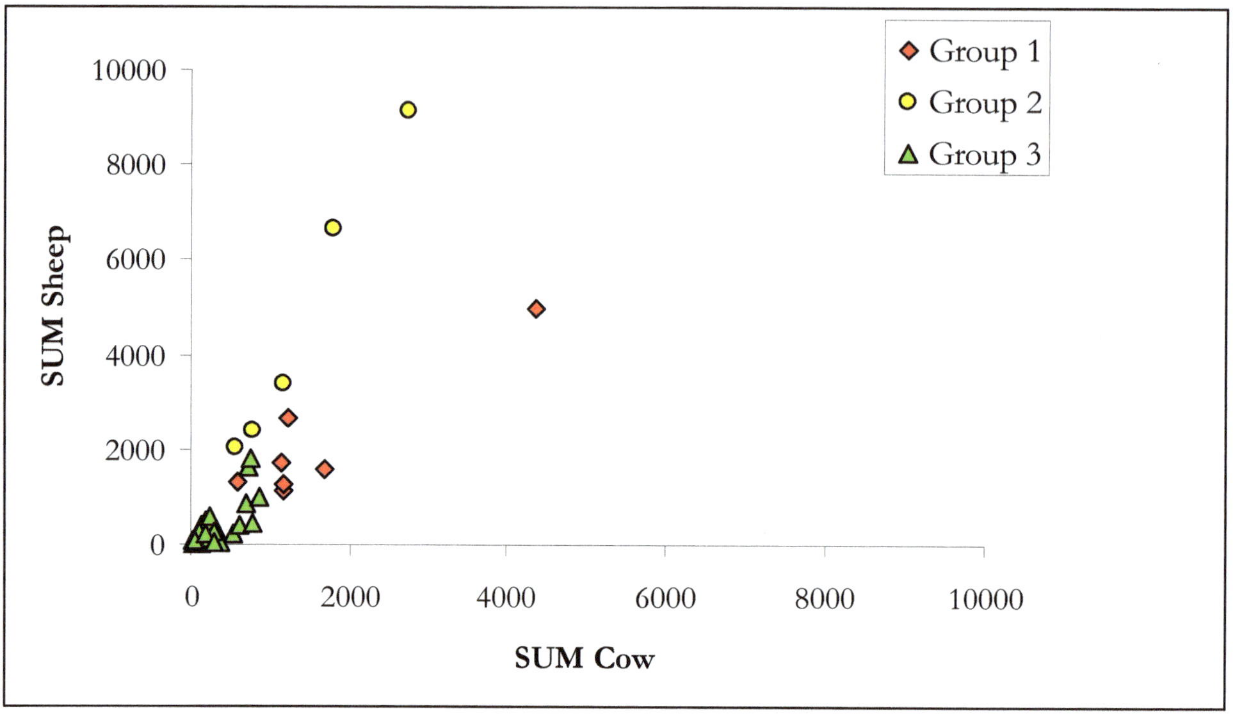

***Figure 38**: The relationship between cow and sheep frequencies separated for the 3 groups identified by PCA*

While the first vector exhibits the *scale* of the assemblages with small sites on the top and large ones at the bottom, the second vector mirrors the *profile* of the sites. If sample size was the only separating factor, all the outlying sites should cluster closer around the y axis. However they spread widely to both sides, which suggests two different types of large Iron Age assemblages. Before the outliers can be regarded in the wider concept of Romanization it has to be established whether they represent changes in human behaviour or just preservation and excavation biases.

Skeletal element representation

A PCA also produces vector loadings for the column variables, in this case the skeletal elements (see table 4). Their first two vectors are plotted in figure 39.

The pig bones have small values and are scattered around the centre of the plot, which indicates that they do not play a major role in the site distributions. This is probably due to the small numbers of pig in the overall assemblage. The only exception is the most abundant pig element, the mandible, which has the highest vector loading with -1.287 (see table 4).

With the exception of the phalanges, all the cow bones plot to the left site of the graph and thus indicate a closer correlation to the group 1 sites which also plot left on their vector representation (compare figure 37). The sheep elements scatter to both sites of the y axis. The sheep tibia and the phalanges show the highest loadings and form the two extremes of the sheep distribution, the tibia to the left and the phalanges to the right. The sheep pelvis and scapula overlap almost completely.

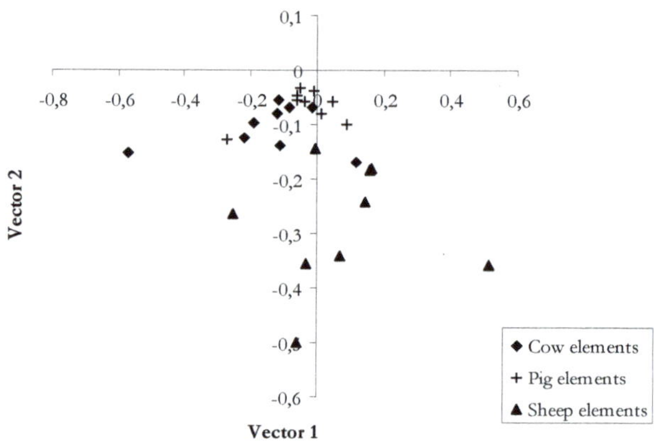

***Figure 39:** Scatter plot for the loadings of the first two components for the skeletal elements*

10.3 DIVISION OF THE DATA SET

Based on these results the assemblage can be split into three parts: "The outliers can then be analyzed separately and the remaining cases analyzed in terms of the shape, strength and direction of the relationship that exists among them. This is an entirely reasonable way to proceed since the outliers are, by definition, not part of the same relationship as the other cases" (Hartwig & Dearing 1979:47). Group 1 contains the 7 sites that cluster on the left, group 2 the 5 sites that cluster to the right and group 3 contains the remaining smaller assemblages (see table 5). Note that the right arm of the distribution only contains the two sites of Maiden Castle and Danebury, but the latter is separated into four time periods. The two time periods given for Balksbury fall into different groups, the very small Early Iron Age sample is in group 3, while the very large Middle Iron Age sample is plotted in group 1.

variable	PC 1	PC 2
c_femur	-0.05613	-0.11602
c_foot	-0.12475	-0.22047
c_humerus	-0.08142	-0.12212
c_mandible	-0.15286	-0.57068
c_pelvis	-0.06918	-0.08456
c_phalanges	-0.16844	0.11817
c_radul	-0.13929	-0.11168
c_scapula	-0.09858	-0.19347
c_tibia	-0.06883	-0.13089
p_femur	-0.03203	-0.05193
p_foot	-0.05918	0.04946
p_humerus	-0.05568	-0.06220
p_mandible	-0.12875	-0.27276
p_pelvis	-0.03786	-0.00930
p_phalanges	-0.09895	0.08809
p_radul	-0.08134	0.01594
p_scapula	-0.05707	-0.03751
p_tibia	-0.04627	-0.05859
s_femur	-0.14315	-0.00538
s_foot	-0.35660	-0.03057
s_humerus	-0.24035	0.14765
s_mandible	-0.50048	-0.06058
s_pelvis	-0.18106	0.16275
s_phalanges	-0.35726	0.51455
s_radul	-0.34153	0.07033
s_scapula	-0.18330	0.16154
s_tibia	-0.26510	-0.25323

Table 4: *Loadings of skeletal elements on the first two principal components, c = cow, p = pig and s = sheep; the accumulated variables are x_foot (metacarpals + metatarsals), x_phalanges (phalange 1, 2 + 3) and x_radul (radius + ulna)*

10.4 RANKING OF SPECIES PROPORTIONS

After the separation of the data set, the ranking of the species abundance is repeated in order to determine whether there are any differences between the three groups. The results are shown in tables 6 and 7 and confirm the results of the graphical analysis. Pig is the least abundant species for all the larger sites and, with the exception of Cat's Water and Ditches, sheep is the most common species. However there seems to be a difference in the proportion of sheep between the outlier groups. Figure 38 shows that the five sites in group 2 have a higher proportion of sheep than all the other sites. Furthermore the average frequencies for cow bones are relatively similar with 1620.57 for group 1 and 1411 for group 2, but for sheep the gap is bigger: 2098.14 for group 1 and 4718.20 for group 2 (see table 6).

Although sheep dominates all assemblages except Ditches and Cat's Water, cattle and pig provided more meat to the diet than a simple bone count suggests. This is due to the weight differences between the two species; Iron Age cattle are thought to have weighed ten times more than sheep (Ryder 1983: 74-83).
The ranked species abundance for group 3 (smaller Slightly fewer than half the sites have a higher percentage of cow than sheep. For two sites, Mount Batten and Skeleton Green, pig is the most abundant species.

10.5 RANK ORDER OF SKELETAL ELEMENTS

Section 9.6 has compared the rank order of skeletal elements between the species in order to identify patterns that might allow suggestions about the use of these animals or the assemblage formation. The next section will investigate whether there are any rank order differences between group 1 and group 2 sites. This analysis might help to explain why they were separated by the PCA and various scatter plots.

Comparison for the overall assemblage

The skeletal elements of the overall assemblage are compared to see whether they show different patterns for group 1 and group 2. The skeletal elements are ranked within each species. The results might be biased by the fact that some of the variables (x_foot, x_phal, x_radul) were created by combining various skeletal elements, which artificially increases the NISP values. This impact becomes visible when the ranked graphs for the overall species distribution are compared to the ones in chapter 9 (figures 22-24). The phalanges for instance ranked low for all three species, but have gained in ranked size now they are added together. However, because the accumulated variables were identically formed for cow, sheep and pig, the rank order should be directly comparable between the species and it is reasonable to assume that differences in rank order are caused by factors other than data preparation.

The two outlier groups show clear differences in the rank order of skeletal elements, especially in the mandibles and phalanges (figures 40-42). While group 1 has more cow and pig mandibles than group 2, the phalanges in both species are much rarer. Interestingly, the sheep shows a different pattern. Here only the phalanges rank significantly lower for group 1 while the mandibles are the most abundant non-accumulated variables in both groups. The sheep tibia ranks higher in group 1.

SITE	PC 1	PC 2	GROUP	SUM NISP
Owslebury	-1677.3	-1103.3	1	11436
Dragonby MIA-LIA	-523.4	-158.3	1	4407
Cat's Water total IA	-239.4	-341.4	1	3542
Puckeridge-Braughing LIA-RB	-76.5	-114.2	1	3382
Balksbury MIA	-164.3	-209.9	1	3037
Ditches LIA-ERB	-6.9	-144.3	1	2706
Rope Lake Hole IA-RB	-9.5	-67.9	1	2043
Danebury MIA-LIA	-2519	508	2	13917
Danebury LIA	-1634.6	426.2	2	9459
Danebury EIA	-658.3	379.1	2	5378
Maiden Castle MIA	-301.5	21.3	2	3471
Danebury MIA	-201.3	283.5	2	3248
Winklebury EIA-MIA	62.5	37	3	2813
Old Down Farm	27.7	57.9	3	2490
Burgh LIA	163.2	-62.5	3	2021
Skeleton Green LIA-RB	243.9	-64.1	3	1875
Brighton Hill South	112.1	-122	3	1723
West Stow MIA-LIA	221.3	-92.2	3	1421
Poundbury total IA	323.4	72.9	3	941
Pennyland MIA	337.3	-16.5	3	845
Meare 1984 LIA	289.8	25.7	3	840
Catcote LIA-RB	352.4	38.4	3	683
Uley Bury	334.1	14.7	3	670
Mount Batten total IA	379.8	52.1	3	639
Dalton Parlours MIA-LIA	309.6	179.4	3	549
Port Seton LIA-ERB	402	-18.6	3	477
Farningham Hill LIA	390.3	28.7	3	457
Abbotstone Down	370.4	12.3	3	450
Grimthorpe IA	401.6	30.3	3	438
Bancroft LIA-ERB	395.2	28.2	3	386
Wavendon Gate	410.7	-1.1	3	379
Bramdean	404.8	52.5	3	302
Ower LIA-RB	419.9	49.4	3	252
Chilbolton Down EIA-MIA	417.5	63.2	3	200
Balksbury EIA	355.8	27.6	3	191
Bishopstone MIA-LIA	434.6	60.2	3	113
Bancroft LBA-EIA	452	67.6	3	58

Table 5: *The 37 British Iron Age sites included in the principal component analysis. The group number describes the location in the plot of the first two PCs. As the Sum NISP values show, the largest sites are placed in group 1 and 2*

SITE	Sum NISP Cow	Sum NISP Pig	Sum NISP Sheep	1	2	3
Group 1						
Balksbury MIA	1140	170	1727	sheep	cow	pig
Cat's Water total IA	1685	264	1593	cow	sheep	pig
Ditches LIA-ERB	1182	384	1140	cow	sheep	pig
Dragonby MIA-LIA	1211	521	2675	sheep	cow	pig
Owslebury	4362	2096	4978	sheep	cow	pig
Puckeridge-Braughing LIA-RB	1155	978	1249	sheep	cow	pig
Rope Lake Hole IA-RB	609	109	1325	sheep	cow	pig
Group 1 mean	1620.57	646	2098.14			
Group 2						
Danebury EIA	1166	835	3377	sheep	cow	pig
Danebury LIA	1790	1011	6658	sheep	cow	pig
Danebury MIA	581	647	2020	sheep	cow	pig
Danebury MIA-LIA	2738	2048	9131	sheep	cow	pig
Maiden Castle EIA-LIA	780	286	2405	sheep	cow	pig
Group 2 mean	1411.00	965.40	4718.20			

***Table 6:** Ranked species abundance (SUM NISP) of sites in group 1 and 2*

GROUP 3 SITES	SUM NISP	SUM Cow	SUM PIG	SUM Sheep	1	2	3
Abbotstone Down	450	221	46	183	cow	sheep	pig
Bancroft LIA-ERB	386	173	50	163	cow	sheep	pig
Farningham Hill LIA	457	239	34	184	cow	sheep	pig
Grimthorpe IA	438	278	39	121	cow	sheep	pig
Pennyland MIA	845	547	56	242	cow	sheep	pig
Port Seton LIA-ERB	477	374	46	57	cow	sheep	pig
Uley Bury	670	297	99	274	cow	sheep	pig
Wavendon Gate	379	312	9	58	cow	sheep	pig
West Stow MIA-LIA	1421	799	155	467	cow	sheep	pig
Bancroft LBA-EIA	58	20	18	20	cow/sheep	pig	
Mount Batten total IA	639	238	264	137	pig	cow	sheep
Skeleton Green LIA-RB	1875	630	860	385	pig	cow	sheep
Bramdean	302	70	52	180	sheep	cow	pig
Brighton Hill South	1723	696	171	856	sheep	cow	pig
Burgh LIA	2021	876	166	979	sheep	cow	pig
Catcote LIA-RB	683	321	36	326	sheep	cow	pig
Chilbolton Down EIA-MIA	200	81	6	113	sheep	cow	pig
Dalton Parlours MIA-LIA	549	126	21	402	sheep	cow	pig
Meare 1984 LIA	840	203	134	503	sheep	cow	pig
Old Down Farm	2490	723	147	1620	sheep	cow	pig
Poundbury total IA	941	255	94	592	sheep	cow	pig
Winklebury EIA-MIA	2813	748	263	1802	sheep	cow	pig
Balksbury EIA	509	190	49	270	sheep	cow	pig
Bishopstone MIA-LIA	113	18	18	77	sheep	cow/pig	
Ower LIA-RB	252	63	92	97	sheep	pig	cow

***Table 7:** Ranked species abundance (SUM NISP) of sites in group 3*

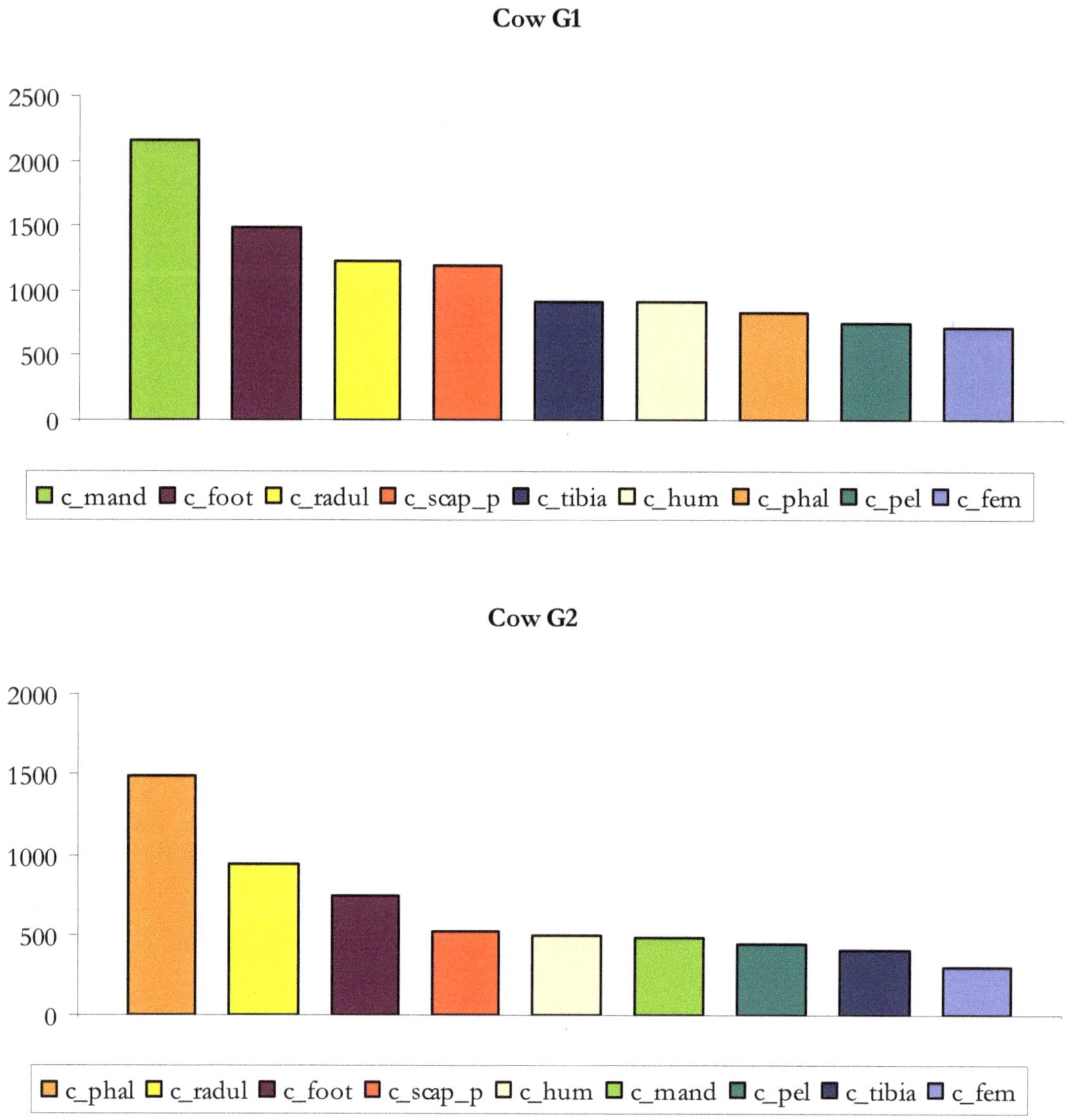

***Figure 40**: The distribution of the skeletal elements of cow for the two outlier groups*

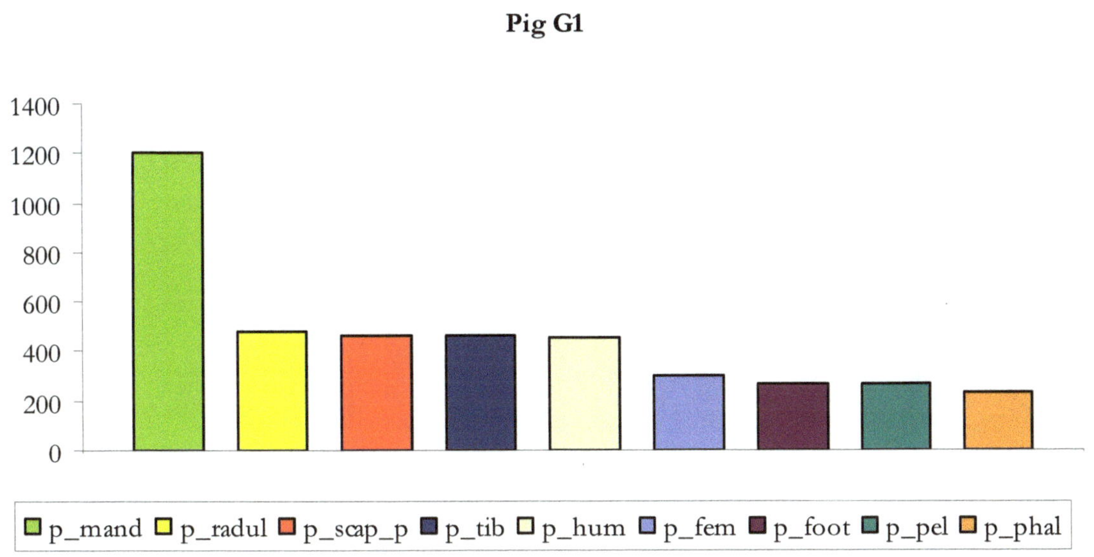

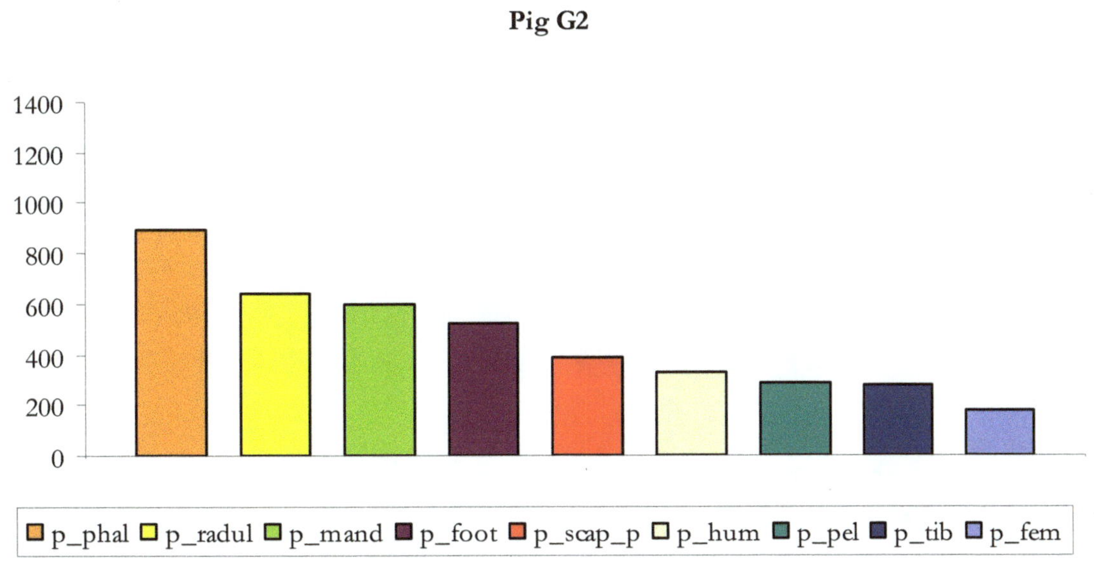

Figure 41: *The distribution of the skeletal elements of pig for the two outlier groups*

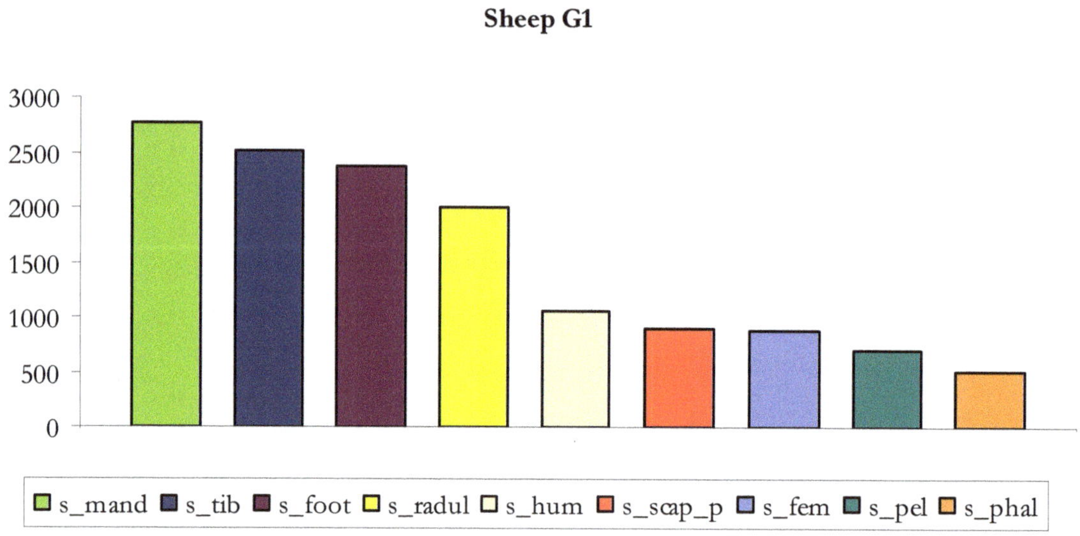

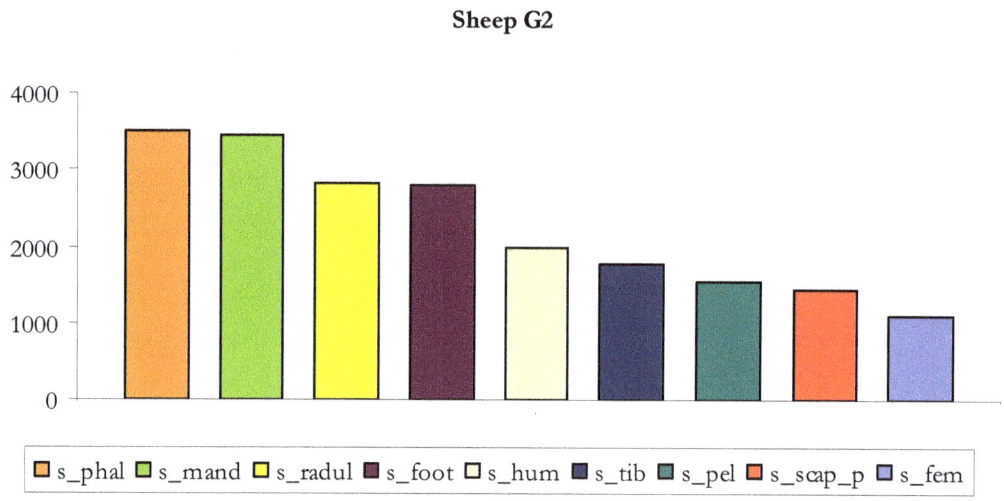

***Figure 42:** The distribution of the skeletal elements of sheep for the two outlier groups*

Spearman rank correlation coefficient

In order to explore the relationship between the rank order of these groups, the Spearman rank correlation coefficient (r_s) is calculated for each species and for all possible group combinations (table 8). This coefficient is a measure of similarity in rank order; it can have any value between – 1.0 and + 1.0. While values close to 0 indicate that there is no correlation between the groups, values close to + 1 suggest a positive correlation and values close to – 1 suggest a negative correlation. A high positive correlation suggests that the rank order in the two groups is the same, implying that both assemblages have the same profile and presumably have been treated the same way.

As in the scatter plots, a weaker correlation (values close to 0) would suggest that the profiles of the two assemblages were independent – the assemblages were treated in different ways. A high negative correlation would suggest that the elements common in one site are rare in another and vice versa. The simplest way this could happen would be for material to be exported from one site to another. If one site contained only low utility elements, for example, and the other high utility elements, that might manifest as a strong negative correlation.

The coefficient calculated for groups 1 and 2 (large assemblages) shows a weak correlation especially for pig and sheep, which indicates that both groups are independent in regard to the rank order of their skeletal elements. There are no appreciable differences between group 1 and the small assemblages in group 3, which raises the question whether sample size is the only factor that separates these two groups. There are no negative correlations.

	G1/ G2	G3/ TOTAL	G1/ TOTAL	G2 / TOTAL	G2/ G3	G1/ G3
PIG	0.100	0.733	0.533	0.867	0.383	0.883
SHEEP	0.233	0.933	0.767	0.667	0.692	0.800
COW	0.433	0.967	0.900	0.767	0.567	0.933

Table 8: *Spearman rank correlation coefficient for the skeletal elements between the groups. The formula is:* $r_s = 1-((6*SRDS)/(N*(N^2-1))$ *with SRDS = the sum of the rank differences squared and N = number of rows in the data, 9 in the present case*

Comparison for the individual sites

Of course this overall pattern might differ from the rank order of the individual sites, in which case the information of the Spearman's rank order coefficient is invalid for a more detailed level of observation. The skeletal elements' rank order is therefore considered for the individual sites in both outlier groups. The results are shown in tables 9-14. As expected, the results are less obvious than for the overall pattern. The rank order of cow elements shows a great variety between the individual group 1 sites and the pattern is not nearly as obvious as the overall distribution (tables 9-11).

Owslebury and Balksbury MIA show the greatest overlap with 5 elements sharing the same rank order. These two sites also confirm the overall rank order of elements with low frequencies for the phalanges and high frequencies for the mandible. The assemblages for Ditches and Rope Lake Hope differ from this overall pattern in that they show high frequencies for both phalanges and mandibles. Dragonby has even higher frequencies for the phalanges than the mandible. With the exception of Puckeridge-Braughing and Dragonby, the mandible is still the highest ranked element for the majority of sites, followed by the foot bones. The femur ranking also overlaps with the overall distribution and shows low frequencies for all 7 sites.

The rankings of sheep and pig elements show both greater similarities to the overall ranking and between the individual group 1 sites. For sheep the phalanges rank low for all 7 sites, which causes the greatest rank order difference to the group 2 sites, where the phalanges are the most abundant elements for all Danebury sites. The sheep mandible shows high frequencies in all group 1 sites, followed by the tibia and the variables foot and radul. While the two accumulated variables also rank high in the cow, the tibia is more abundant for sheep than for cow. Although Balksbury MIA showed the greatest similarities to Owslebury in the cow element ranking, it is almost identical to Rope Lake Hole in its ranking of sheep bones. Cat's Water shows the same ranking as these two sites in 5 elements. For the sheep, Owslebury shows the greatest overlap with Ditches in 4 elements followed by Puckeridge-Braughing with 3 elements.

The clustering of sites is again different for the pig bones. Here Balksbury MIA and Dragonby overlap in the rank order of 7 elements and Ditches and Rope Lake Hope for 5 elements. All group 1 sites confirm the overall pattern with the mandible as the most abundant element and the femur and phalanges showing low frequencies.

Compared to the first outlier group, the group 2 sites have a much greater overlap in rank order (tables 12-14). This is certainly due to the fact that the group only contains two sites and the element distribution is very similar for all four Danebury periods. However the Danebury assemblages and Maiden Castle show significant rank order differences for all three species. Maiden Castle has fewer sheep and pig phalanges. The femur is the least abundant element while the phalanges rank high for all three species at both sites. The cow mandible is less frequent than the mandible for the other two species.

The accumulation of individual group 1 sites' frequencies had the most distorting effect for the cow elements. Here the overall distribution suggests a pattern that is not repeated in some sites. But generally the greatest rank order differences between the groups remain in the mandibles and phalanges: for most of the group 1 sites the mandible ranks high in all three species, while the phalanges fall into lower ranks. The only exceptions are Ditches and Rope Lake Hole with high phalanges frequencies for the cow bones and Puckeridge-Braughing with an exceptionally low frequency for the cow mandibles. For all the group 2 sites the femur ranks very low for all three species. The mandible ranks significantly lower for the cow bones, while the phalanges are within the highest rank elements for all three species. The rank order of Maiden Castle differs from the four Danebury periods.

None of the individual sites overlaps with any of the other sites for more than one species and a clustering of sites within the groups is thus not possible based on the rank order of skeletal elements.

COW	mandible	tibia	radul	humerus	femur	Scapula	pelvis	foot	Phalanges
Balksbury MIA	9	4	5	8	2	6	3	7	1
Owslebury	9	4	6	5	2	7	3	8	1
Ditches	9	3	4	1	2	8	5	6	7
Rope Lake Hole	9	2	6	4	1	5	4	8	7
Puckeridge	2	6	8	4	1	7	3	9	5
Dragonby MIA-LIA	6	4	9	2	1	5	3	7	8
Cat's Water	8	5	7	6	4	3	1	9	2

Table 9: *Rank order of cow skeletal elements for the group 1 sites. 9=most abundant element, 1=least abundant element*

SHEEP	mandible	tibia	radul	humerus	femur	Scapula	pelvis	foot	Phalanges
Puckeridge	9	6	7	4	3	8	2	5	1
Owslebury	9	8	6	4	5	1	2	7	3
Ditches	9	8	6	4	2	7	5	3	1
Rope Lake Hole	7	8	6	5	4	3	3	9	1
Balksbury MIA	7	8	6	5	4	3	2	9	1
Cat's Water	6	8	7	5	3	4	2	9	1
Dragonby MIA-LIA	8	9	6	5	3	4	2	7	1

Table 10: *Rank order of sheep skeletal elements for the group 1 sites*

PIG	mandible	tibia	radul	humerus	femur	Scapula	pelvis	foot	Phalanges
Owslebury	9	7	5	8	3	6	2	1	4
Puckeridge	9	7	8	4	2	6	4	5	1
Balksbury MIA	9	5	6	7	2	8	4	3	2
Dragonby MIA-LIA	9	5	6	7	4	8	1	3	2
Cat's Water	9	8	6	5	4	7	2	3	1
Ditches	9	3	8	5	1	6	5	7	3
Rope Lake Hole	9	5	8	5	1	8	2	6	3

Table 11: *Rank order of pig skeletal elements for the group 1 sites*

COW	femur	phalanges	radul	foot	humerus	mandible	tibia	scapula	Pelvis
Danebury LIA	1	9	8	7	3	4	5	6	2
Danebury MIA	1	9	8	7	4	6	2	5	3
Danebury MIA-LIA	1	9	8	7	5	2	3	6	4
Danebury EIA	1	9	7	8	5	4	2	3	6
Maiden Castle	3	7	8	9	5	6	3	4	1

Table 12: *Rank order of cow skeletal elements for the group 2 sites. 9=most abundant element, 1=least abundant element*

SHEEP	femur	phalanges	radul	foot	humerus	mandible	tibia	scapula	Pelvis
Danebury MIA	1	9	6	7	5	8	3	2	4
Danebury MIA-LIA	1	8	6	7	5	9	3	4	2
Danebury LIA	1	8	7	6	5	9	4	2	3
Danebury EIA	1	9	8	6	5	7	4	2	3
Maiden Castle	4	3	8	9	5	6	7	1	2

Table 13: *Rank order of sheep skeletal elements for the group 2 sites*

PIG	femur	phalanges	radul	foot	humerus	mandible	tibia	scapula	Pelvis
Danebury MIA	1	9	8	7	5	6	2	3	4
Danebury EIA	1	9	8	7	2	6	4	5	3
Danebury LIA	1	9	7	6	2	8	4	5	3
Danebury MIA-LIA	1	9	8	6	4	7	2	5	3
Maiden Castle	1	5	9	3	6	7	2	9	4

Table 14: *Rank order of pig skeletal elements for the group 2 sites*

10.6 SCATTER PLOTS FOR SKELETAL ELEMENTS WITH RANK ORDER DIFFERENCES

The last section suggests that the elements with the highest rank order differences (pig and cow mandibles and phalanges, sheep tibia and phalanges) might be responsible for the separation of group 1 and group 2. This hypothesis is tested visually by identifying the outlier groups on various scatter plots that showed a bifurcation in the exploratory data analysis. If these elements are indeed causing the pattern, frequencies for the group 1 and group 2 sites should plot along different regression lines.

The scatter plot for the pig and cow mandible does not separate the two outlier groups when plotted against the cow mandible (figure 43). The relationship between sheep tibia and pig mandible (figure 44) also shows no distinctive pattern dividing groups 1 and 2, even though both elements are relatively more frequent in group 1. The ranking of these elements is overwritten by their absolute frequencies and does not seem to be an essential attribute of the group formation. One could argue that both are very robust elements. However both have different utility indices and are unlikely to have been treated as one unit at the site. It is therefore not surprising to find no clear correlation between the two. This shows that the initial assumption is not valid for some of the elements which, in fact, show a rather random pattern.

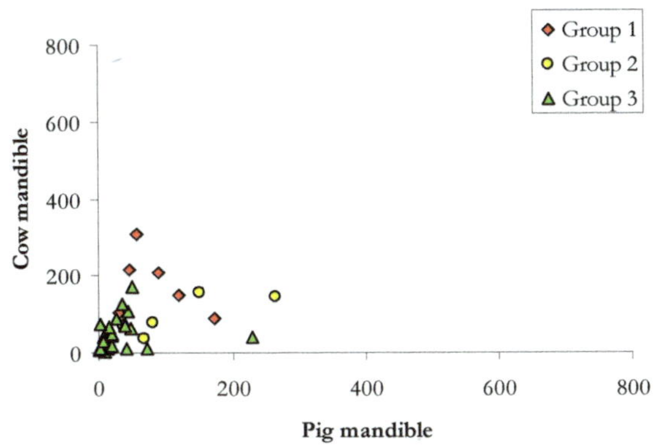

Figure 43: *The relationship between cow and pig mandibles for all three Iron Age groups. The graph shows that the statistical separation of sites is not confirmed visually for these two elements.*

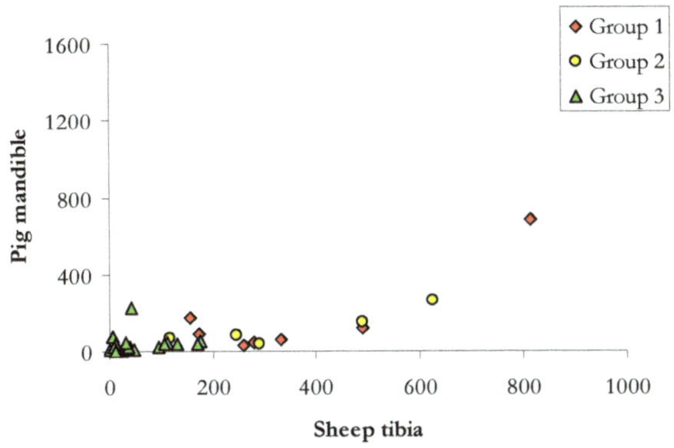

Figure 44: *The relationship between sheep tibia and pig mandibles for all three Iron Age groups*

Although the pig mandible is the most abundant element in group 1 (see figure 41) its frequencies are independent from the group 2 sites. Because pig is the least abundant species for all outlier groups, the cow mandibles have higher absolute frequencies even if they rank significantly lower. Only one group 2 site (Puckeridge-B.) has absolutely more pig than cow mandibles. The only group 3 site that always falls within the outlier groups when the pig mandibles are plotted is Skeleton Green, one of the few sites where pig is the most abundant species.

However figure 45 shows a clear bifurcation between the sheep phalanges and the cow mandibles, also two elements with low utility indices and no logical connection. They are distributed evenly throughout each group and the plot is clearly separated into three groups. All group 2 sites exhibit exceptionally low frequencies of sheep phalanges and thus confirm the overall pattern in the rank order of skeletal elements.

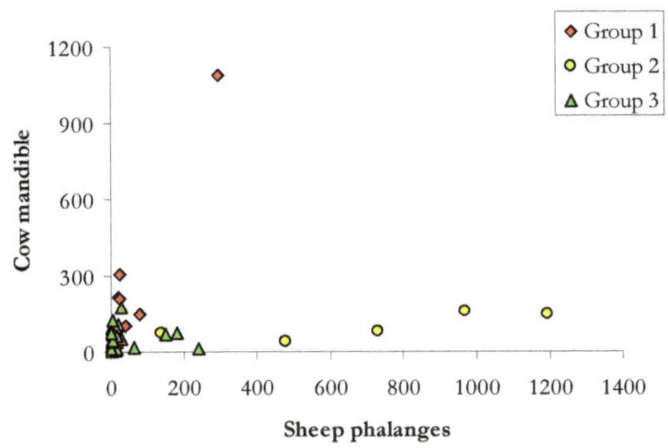

Figure 45: *The relationship between sheep phalanges and cow mandibles for all three Iron Age groups*

The pig mandible also shows a clear bifurcation when plotted against the sheep phalanges (figure 46). All group 2 sites contain relatively more sheep phalanges while the group 1 sites have relatively more pig mandibles. When the pig mandible is plotted against the cow phalanges the bifurcation is less obvious but still there (figure 47). For element frequencies below 100 the relationship between elements is similar for all groups, but the larger numbers clearly show a higher number of pig mandibles for the group 1 sites. The pattern repeats itself when the cow mandibles are plotted against the phalanges of all three species (not shown here).

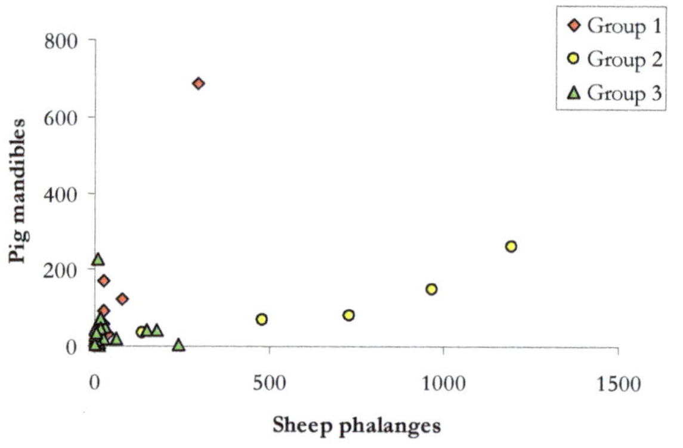

Figure 46: The relationship between sheep phalanges and pig mandibles for all three Iron Age groups

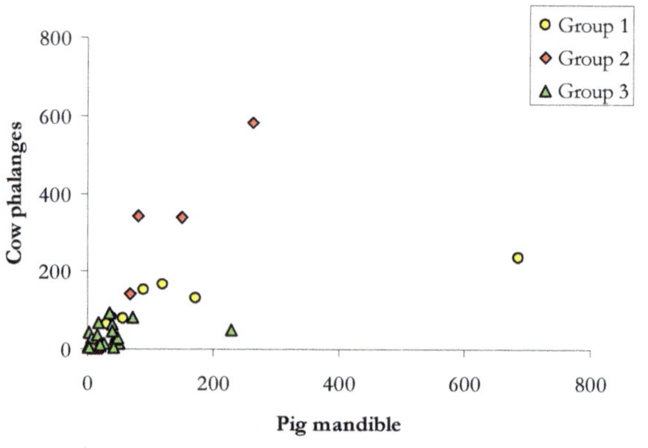

Figure 47: The relationship between cow phalanges and pig mandibles for all three Iron Age groups

Plotting the sheep tibia shows interesting patterns because Maiden Castle, the second site in group 2, sometimes falls within the group 1 sites rather than its own group (figure 48). For this large Iron Age hillfort the tibia falls in rank 7 for cow and pig and in rank 3 for sheep, which is closer to the group 1 distribution. When plotted against the sheep phalanges the group 2 sites show more sheep phalanges than sheep tibia, except for Maiden Castle.

This analysis has shown that not all elements with rank order differences seem to play a part in the outlier constitution.

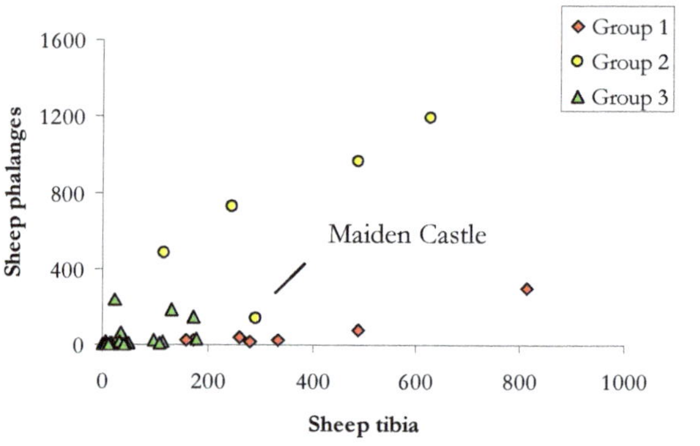

Figure 48: The relationship between sheep phalanges and sheep tibia for all three Iron Age groups.

10.7 CONCLUSIONS

The PCA has separated the British Iron Age faunal assemblage into three groups. The differences between groups 1 and 2 (large assemblages) and the remainder of the data are correlated with sample size. However the two outlier groups show differences both in the rank order of skeletal elements and the species proportions. This indicates that their profiles are different.

Group 1 contains the sites Owslebury, Dragonby MIA-LIA, Cat's Water total IA, Puckeridge-Braughing LIA-RB, Balksbury MIA, Ditches LIA-ERB and Rope Lake Hole IA-RB. Compared to group 2, which contains the 4 periods for Danebury as well as the faunal remains found in Maiden Castle, the group 1 sites contain generally fewer sheep phalanges and more sheep tibia. The group 1 sites have higher frequencies for pig and cow mandibles and only very low values for the phalanges. There are proportionally more sheep bones in group 2 than in group 1. Not every skeletal element that has shown an obvious rank order difference between the outlier groups shows a different composition between the groups. A clear separation of the two outlier groups is visible, both in the ranking of elements as well as in a visual representation between the cow mandible and the phalanges of all three species and the pig mandible and the phalanges of all three species. The group 2 sites contain proportionally more sheep bones than the group 1 sites, although sheep dominates almost all the assemblages.

11. TEST FOR PRESERVATION AND RECOVERY BIASES

Considering the dataset and the general nature of faunal remains, taphonomic processes or differential recovery are the most likely explanations for the observed differences between the outlier groups. Many archaeologists have argued that any reasonable analysis of behavioural change is only possible after these potential biases are excluded as an explanatory factor for the observed pattern (Binford 1981, Hambleton 1999). However it has also been shown here that the archaeological record never allows us to exclude all potential biases or unpick the confounding problems (see also Winder 1996, 1997).

Before the data is interpreted this study will try to identify the potential biases that might have acted on the Iron Age remains and investigate how they might have changed the original assemblage. It is not assumed that all the biases can be fully recognized, but the observed pattern will be interpreted afterwards. The archaeological record will never provide us with more accurate data and surrendering to its difficulties would lead to the exclusion of all empirical studies.

11.1 MANDIBLE AND PHALANGES AS INDICATORS

Due to the lack of data - MNI values or sample weights are unavailable for most sites - common methods to test for the fragmentation or preservation of the bones cannot be applied. However the skeletal elements that constitute the main differences between both outliers can give an insight into the processes that acted on the assemblages in two different ways:

1) Payne (1972a) has demonstrated that sieving increases the recovery of small bones from larger animals. Good preservation also leads to higher survival rates of small or less dense bones. However when the preservation is bad, small and fragile bones are likely to be destroyed and even if the recovery techniques are very precise it is often impossible to recover many small bones in such conditions. Therefore high frequencies of small *phalanges* can be used as an indicator for good preservation and/or recovery. Based on the small bones alone, it is difficult to determine whether their frequencies represent the impact of preservation or recovery. This provides an example for the problems arising out of the dynamic, multi-temporal nature of the archaeological record. The impact of long-term taphonomic processes and short-term excavation are blurred into the static bone counts on which this analysis is based.

2) The second element that can give evidence about the state of preservation is the *mandible*, not because of its size but because it contains the teeth and thus the most robust parts of a skeleton. Teeth are likely to survive even when most other bones have disappeared. If the taphonomic processes acting on the remains are very strong, loose teeth can become separated from the mandible and appear in large numbers in the assemblage. The number of loose teeth (bad preservation) in relation to surviving mandibles (good preservation) has been used as an indicator for the state of preservation. In Balksbury relatively more loose teeth were found for sheep than for cattle (cow: teeth: 174/mandible: 205; sheep: teeth: 362/mandible: 273) which shows that the sheep assemblage was more affected by taphonomic processes than the cow bones (Maltby 1995).

Section 8.1 has discussed how the fragmentation of particular bones can bias the NISP counts, because they simulate a higher presence of bones than the number that actually entered the archaeological record. One complete mandible will be counted as one element but a fragmented bone or individual teeth will give higher frequencies. Most of the outlier site reports explicitly exclude individual teeth from the mandible counts and it is thus unlikely that the mandible counts have been artificially increased by the inclusion of individual teeth. However the shape of the mandible is very characteristic and it is easy to identify even when it is fragmented. It is therefore more likely to be identified than other less indisputable elements and thus more likely to enter the element counts of a site. Exceptionally high mandible counts are therefore likely to represent fragmented bones and thus bad preservation. Low mandible counts, on the other hand, indicate good recovery of less fragmented bones. These two hypotheses are visualised in table 15. The table demonstrates the difficulties in interpreting archaeological data due to the confounding of causes. Low frequencies for the phalanges may be the effect of partial recovery as well as bad preservation. We do not know how this dialectic is resolved in any one data set.

<table>
<tr><td colspan="2"></td><td colspan="2">PRESERVATION</td></tr>
<tr><td colspan="2"></td><td>GOOD</td><td>BAD</td></tr>
<tr><td rowspan="2">RECOVERY</td><td>GOOD</td><td>Phalanges ↑
Mandible ↓</td><td>Phalanges ↓
Mandibles ↑</td></tr>
<tr><td>PARTIAL</td><td>Phalanges ↓</td><td>Teeth ↑
Phalanges ↓</td></tr>
</table>

Table 15: *Mandible and phalanges frequencies as indicators for the preservation and quality of recovery techniques for faunal remains. ↑: high frequencies ↓: low frequencies*

According to the hypothesis the data suggest good preservation and/or recovery for the group 2 sites (low frequencies for the mandibles and a high percentage of phalanges) and bad preservation or partial recovery for the group 1 sites (low frequencies for the phalanges and a high percentage of mandibles). If the overall skeletal element distribution is considered (figures 40-42) the hypotheses fit better for the cow and pig than for the sheep assemblages. For the group 2 sites the high frequencies of phalanges for all three species, as well as the low counts of cow and pig mandibles, indicate good preservation. This is in consistence with the site reports for Danebury and Maiden Castle. Both sites are located on the chalklands of Wessex, ground that provides good conditions for bone preservation (Maltby 1994). Only a few of the Danebury deposits were sieved and the monitoring of the results suggests that - despite their already high frequencies - small bones and teeth are actually underrepresented for the larger animals, cow, sheep and pig (Grant 1984a: 496). The fact that many of the bones recovered at Danebury are from young lambs and calves and thus rather fragile is a clear indication for the excellent preservation at the site. "No other samples from the region exhibit neonatal/infant mortality levels as high as at Danebury" (Hambleton 1999: 82). About one third of the animals died in their first year of life and even complete skeletons of neonatal lambs have been recovered. However the sheep and, to a lesser extent, the pig assemblages for these two sites show a high distribution of mandibles, which cannot really be explained by partial preservation. A detailed rank order of skeletal elements (tables 12-14) for each site has shown that this pattern is consistent for all group 2 sites and not caused by one outlier with exceptionally high mandible frequencies.

For the overall skeletal element ranking in group 1 the mandible is the most abundant element for all three species and phalanges the least abundant for pig and sheep (see figures 40-42). This summary suggests bad preservation and/or partial recovery for this group. The individual site reports generally confirm this assumption. As in group 2 the hypotheses fit the sheep and pig bones better than the cow. This is probably due to size differences between the three species. Cow bones are bigger and therefore more robust than sheep bones and small cow elements are therefore more likely to survive when they get into the ground. Some sites deviate from the general pattern: the Puckeridge-Braughing assemblage contains very few cow mandibles and Ditches, Rope Lake Hope and Dragonby have high phalanges frequencies which shift this cow element to place 3 in the overall ranking. Both aspects indicate good rather than bad preservation. However this deviation might be explained with a higher survival rate of the larger and more robust cow elements. The site report for Rope Lake Hole supports this assumption as it suggests a recovery bias that led to a less efficient retrieval of sheep carpals, tarsals and phalanges. Thus taphonomy and excavation techniques acted more strongly on the sheep than the cow elements. The same observation has been made for Balksbury, where the recovery of smaller sheep bones was less efficient than for cattle (Maltby 1995). In this case however the bias does not show in the rank order of skeletal elements; the phalanges are the least frequent for both species.

Generally the animal bones recovered at Rope Lake Hole were well preserved but highly fragmented (Coy 1987). The site report for Dragonby states that the animal bones were well preserved although many were fragmented and there was a scarcity of very small bones. "Some virtually complete skeletons lacked many of the small bones, and there is an almost total absence of small mammals of hedgehog, squirrel and weasel size" (Harman 1996: 141). While sieving for plant remains some small bones were recovered which indicates that the lack of phalanges for the sheep and pig assemblages in Dragonby is caused by recovery techniques rather than taphonomic processes. Due to their size, cow phalanges are more likely to be recovered even when small bones are missed.

Recovery at Puckeridge-Braughing was also biased towards smaller bones which are underrepresented in the faunal assemblages. A trial sieving confirmed that the excavation missed many of the small bones due to recovery techniques rather than bad preservation. On the other hand, the high number of loose teeth found at the site also supports the assumption of bad preservation (Filfield 1988).

The faunal assemblage of Owslebury has been influenced by taphonomic processes, the majority of bones were affected by erosion, weathering and scavenging. The effects of these factors are stronger on the upper (later) layers, which reduces the information that can be gained from an intra-site comparison. The impact of taphonomic processes also differed from species to species. Cow bones were better preserved than sheep and pig, and the sheep bones were most severely affected by taphonomic factors (Maltby 1987).

Roughly speaking, the group 2 sites show good preservation while the group 1 sites show high fragmentation and/or differential recovery, but the evidence is not as clear-cut for the group 1 assemblages (table 16).

Therefore another means is chosen to test for bone preservation in the outlier groups. The proportion of small bones in relation to the overall species distribution is determined by dividing each skeletal element by the SUM NISP value of the corresponding species. Proportionally higher frequencies of the skeletal element will produce higher results; lower frequencies will produce lower results. According to the hypotheses, higher values in the overall assemblage/mandible relation would be expected for the poorly preserved assemblages in group 1, while the good preservation of the group 2 sites would produce relatively higher values for the phalanges. The results are shown in table 17.

G	SITE	Species	Phal	Mand	Preservation	Frag	Recovery	Kill-off pattern	Husbandry
2	Danebury	Cow	↑	↔	good			secondary	
		Sheep	↑	↑	good			young & secondary	lambing at site
2	Maiden Castle	Cow	↑	↑	good			secondary	
		Sheep	↓	↔	good			secondary	lambing at site
1	Owslebury	Cow	↓	↑	good			**meat,** secondary	
		Sheep	↓	↑	bad			**meat**	
1	Ditches	Cow	↑	↑		high		secondary	
		Sheep	↓	↑				**meat,** secondary	**likely outside**
1	Rope Lake Hole	Cow	↑	↑	good	high	less affected	?	
		Sheep	↓	↑	good	high	partial	**meat**, secondary	
1	Dragonby	Cow	↑	↔	good	high	less affected	secondary	calving at site
		Sheep	↓	↑	good	high	partial	young, secondary & **meat**	**outside**
1	Cat's Water	Cow	↑	↓				secondary	**likely outside**
		Sheep	↔	↓				**meat**	
1	Puckeridge	Cow	↔	↓	bad		less affected	young & secondary	
		Sheep	↓	↑	bad		partial	secondary	
1	Balksbury	Cow	↓	↑	good		less affected	secondary	
		Sheep	↓	↑	bad		partial	young & secondary	lambing at site

Table 16: *Characteristics pointing towards preservation, recovery and behavioural practices at the outlier sites*
↑: high frequencies, ↓: low frequencies ↔ medium frequencies, information on Danebury as average of 4 time periods. ***young****: lambs or calves, natural losses,* ***meat****: ideal age for culling for meat,* ***secondary****: animals older than the ideal age for meat production & used for secondary products or traction*
?: unknown

While the sheep phalanges are generally less frequent in the group 1 outliers, the cow phalanges are more frequent which is probably due to the size differences between both species. This factor also appears to influence recovery, which tends to be less efficient for sheep than for cow bones.

Age at the time of death can also influence the survival rate of bones in the archaeological record which is why the kill-off pattern might explain the differences between the outliers. However the individual site reports reveal that the kill-off patterns for sheep divide the assemblages across the group lines; while sheep in Dragonby, Ditches, Owslebury and Rope Lake Hole were probably exploited for meat, the other sites used them mainly for secondary products and thus killed the animals off later in life. However age at death does not overlap with the overall state of preservation in either of the outlier groups.

With the exception of cow bones at Puckeridge-Braughing and Dragonby all the group 1 sites have higher proportions of mandibles than phalanges. The pattern is most dominant in Owslebury; its mandibles account for more than 30% of the entire pig assemblage. As expected, the group 2 sites show the opposite pattern, with higher frequency counts for phalanges than mandibles. The only exceptions are the sheep elements, where mandibles are more abundant for Danebury LIA, Danebury MIA-LIA and Maiden Castle and the cow mandible frequency in the Maiden Castle assemblage.

These results suggest that the variation between the groups is caused by differential preservation and/or recovery techniques rather than behavioural differences. In regard to the initial questions of behavioural differences between British Iron Age assemblages this result is disappointing as it does not allow the detection of innovation.

GROUP	SITE	c_mand/ SUM	c_phal/ SUM	p_mand/ SUM	p_phal/ SUM	s_mand/ SUM	s_phal/ SUM
1	Balksbury MIA	0.190	0.017	0.271	0.000	0.159	0.010
1	Cat's Water total IA	0.183	0.046	0.216	0.004	0.159	0.016
1	Ditches LIA-ERB	0.175	0.127	0.234	0.057	0.181	0.020
1	Dragonby MIA-LIA	0.124	0.135	0.230	0.054	0.170	0.030
1	Owslebury	0.249	0.054	0.327	0.072	0.231	0.059
1	Puckeridge-Braughing LIA-RB	0.076	0.113	0.175	0.027	0.167	0.020
1	Rope Lake Hole IA-RB	0.169	0.102	0.275	0.055	0.171	0.029
2	Danebury EIA	0.066	0.292	0.096	0.234	0.113	0.216
2	Danebury LIA	0.087	0.188	0.147	0.159	0.155	0.145
2	Danebury MIA	0.067	0.241	0.104	0.249	0.119	0.238
2	Danebury MIA-LIA	0.053	0.213	0.128	0.173	0.165	0.131
2	Maiden Castle EIA-LIA	0.091	0.109	0.126	0.073	0.119	0.057

Table 17: *The relationship of phalanges and mandibles to the overall species abundance*
c_x = cow, p_x = pig, s_x = sheep

However differential preservation and recovery biases may not account for all the observed differences in species distribution between the sites.

11.2 SECOND PCA

To gain absolute certainty about the outlier formation another Principal Component Analysis is undertaken; this time excluding the bones that have been used as indicators for bad preservation and recovery biases (mandibles and phalanges). If these elements are the main determinant of the differences between the outlier groups, the new PCA should show considerably less scattering or no outliers at all. The results are shown in tables 18 & 19 and figures 49 & 50)

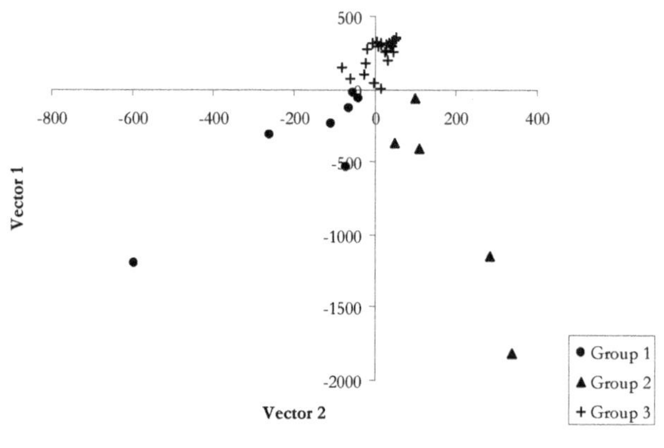

Figure 49: *Relationship of the first two site vectors for 37 British Iron Age sites and 7 skeletal elements for the species cow, sheep and pig, based on a variance-covariance matrix.*

A graphical representation of the first two vectors of this second PCA shows basically the same pattern as the first one, all outlier sites plot in the same quadrant (figure 49).

This means that even though differential preservation and recovery bias seem to explain a great amount of the difference between the two groups, they cannot be the only factor. The remaining differences are likely to represent human impact on the assemblages.

The graphical representation of the vector loadings for the skeletal elements (figure 50) shows a clearer picture than the one of the first PCA. All pig bones cluster around 0 and are thus not significant for the separation of the groups. With the exception of the sheep tibia, the elements of other two domesticates are separated by the Y axis; all the cow elements have negative values on the second vector while, with the exception of the tibia, all the sheep elements have positive values (table 19). As with the previous PCA, the sheep tibia is identified as an outlier. The histograms in chapter 11 have indicated that the group 1 sites contain higher frequencies of sheep tibia than the group 2 sites. This could be related to tool manufacture; sheep tibia and metapodials were used in preference to other bones for the production of bone objects at Danebury (Grant 1984a: 501), which could explain why this element is relatively rare in the species counts. However there is evidence that the sheep tibia has also been used for tool production at some of the group 1 sites.

The differential tibia frequencies could also be attributed to preservational biases. The group 1 sites represent less well-preserved sites or excavation biases. The tibia is one of the most robust long bones in sheep (Brain 1981) and thus more likely to survive than some of the other, more fragile bones. This might explain while it is more frequent in the group 1 sites. In the better-preserved group 2 sites the distribution of skeletal elements is more even.

The overall pattern confirms the previous observation that the group 2 sites contain more sheep than cow while the group 1 sites contain more cow than sheep. King (1978) has shown that the majority of Roman assemblages exhibit a higher percentage of cow rather than sheep which seems to indicate a behavioural shift away from the Iron Age economy.

SITE	GROUP	VECTOR 1	VECTOR 2
Balksbury MIA	1	-234.1	-109.8
Cat's Water total IA	1	-314	-261
Ditches LIA-ERB	1	-23.5	-52.6
Dragonby MIA-LIA	1	-532.3	-71.5
Owslebury	1	-1200.3	-597.3
Puckeridge-Braughing LIA-RB	1	-129.4	-64.4
Rope Lake Hole IA-RB	1	-61.1	-40.5
Danebury EIA	2	-406.1	107
Danebury LIA	2	-1148.3	283.7
Danebury MIA	2	-63.3	98.7
Danebury MIA-LIA	2	-1813	338.1
Maiden Castle EIA-LIA	2	-367.6	47.4
Abbotstone Down	3	294.5	5.7
Balksbury EIA	3	261.3	22.8
Bancroft LBA-EIA	3	357	49.8
Bancroft LIA-ERB	3	306.6	26.1
Bishopstone MIA-LIA	3	342.6	51.5
Bramdean	3	338.7	42.4
Brighton Hill South	3	75.4	-62.2
Burgh LIA	3	98.8	-25.9
Catcote LIA-RB	3	266.8	26.6
Chilbolton Down EIA-MIA	3	330.1	40.6
Dalton Parlours MIA-LIA	3	296.5	39.7
Farningham Hill LIA	3	307.6	14
Grimthorpe IA	3	316.6	12.7
Meare 1984 LIA	3	198.1	29.9
Mount Batten total IA	3	315	34
Old Down Farm	3	9.3	12.2
Ower LIA-RB	3	330.7	47.1
Pennyland MIA	3	273	-20.9
Port Seton LIA-ERB	3	318.3	-5.1
Poundbury total IA	3	256.7	45.3
Skeleton Green LIA-RB	3	184.9	-23.7
Uley Bury	3	293.1	38.4
Wavendon Gate	3	322.1	4
West Stow MIA-LIA	3	151.9	-80.8
Winklebury EIA-MIA	3	47.6	-1,9

Table 18: *Vector loadings of the Iron Age sites for the second PCA*

ELEMENT	VECTOR 1	VECTOR 2
c_fem	-0.07828	-0.19936
c_foot	-0.17299	-0.38637
c_hum	-0.11219	-0.20724
c_pel	-0.09412	-0.13741
c_radul	-0.1901	-0.17498
c_scap	-0.13465	-0.31736
c_tibia	-0.09591	-0.22372
p_fem	-0.04287	-0.08753
p_foot	-0.07739	0.09417
p_hum	-0.07469	-0.09596
p_pel	-0.05019	-0.00793
p_radul	-0.10786	0.04214
p_scap	-0.07703	-0.05163
p_tib	-0.06262	-0.09525
s_fem	-0.19331	0.02651
s_foot	-0.48592	0.01448
s_hum	-0.32086	0.32208
s_pel	-0.23936	0.33896
s_radul	-0.4626	0.19538
s_scap	-0.24398	0.33984
s_tibia	-0.36656	-0.394

Table 19: *Vector loadings of skeletal elements for the second PCA*
c_x = cow, p_x = pig, s_x = sheep

This pattern could also be an artefact of geography, because the chalk lands produce only medium quality grass and thus provide better conditions for sheep than cattle husbandry. Now it has been established that human behaviour is responsible for some of the group differences, a thorough investigation is needed in order to identify which processes are most likely to have caused this pattern.

11.3 CONCLUSION

These analyses of the two outlier groups have shown that differential preservation and/or recovery techniques seem to play an important role in the formation of these groups. Generally speaking the group 2 sites show good, and the group 1 sites bad, preservation and/or partial recovery. This result confirms once more the impact of bone fragmentation or recovery techniques on the frequency counts. It clearly demonstrates the need for a consistent way of recording and quantification in archaeology if we want to proceed with intra-site comparison. However the second Principal Component Analysis showed that the differences between the two outlier groups remained even after the skeletal elements linked to recovery and preservation differences were removed. Therefore partial preservation and recovery may not be the only cause of the inter-group variation. The following chapter will try to investigate which behaviour might be responsible in order to connect this case study to the conceptual model and the processes of Romanization.

Figure 50: *Scatter plot of the vector loadings for the 21 skeletal elements of the second PCA*

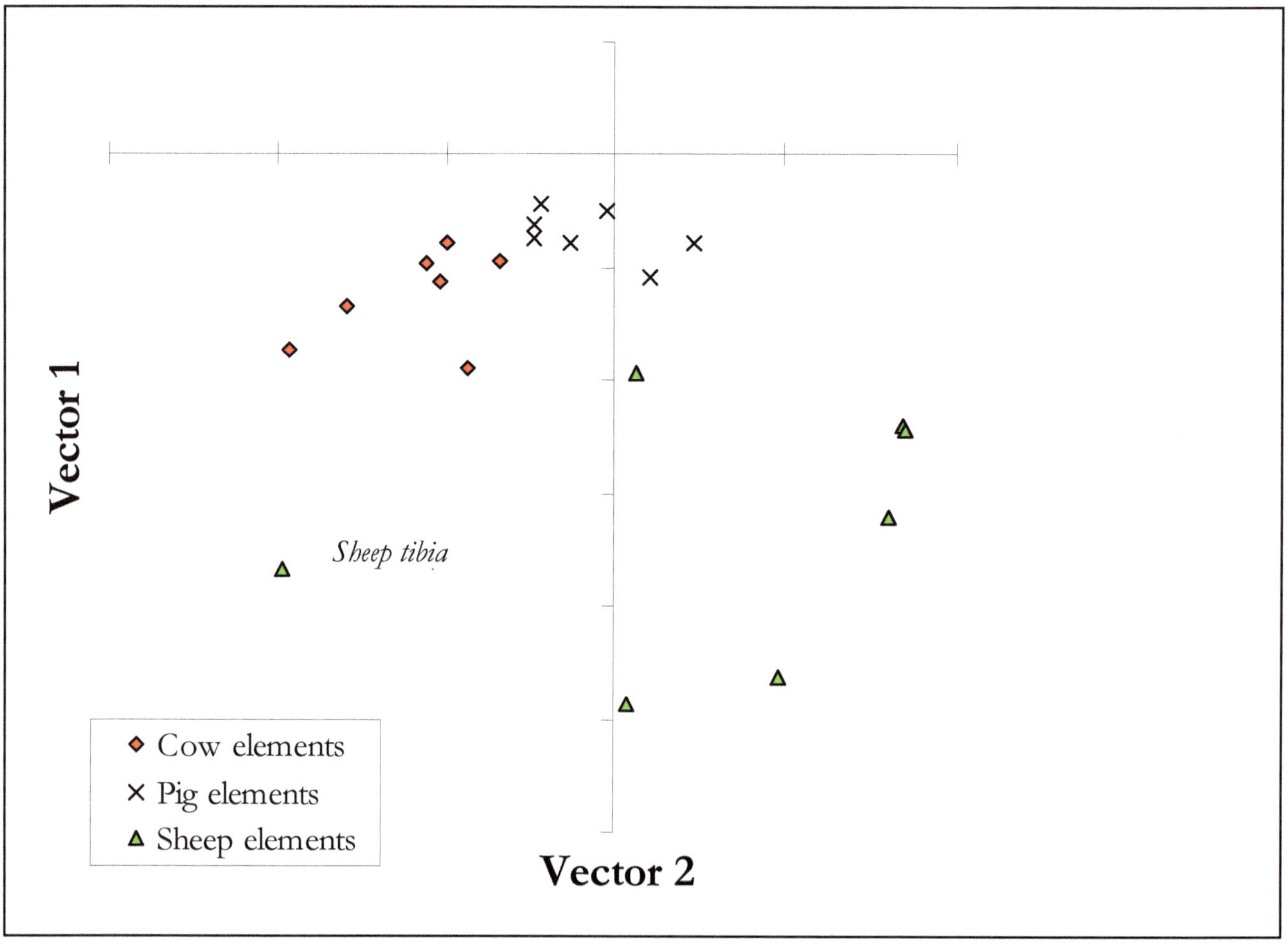

12. DATA DISCUSSION

The Romans occupied Gaul about 150 years before they invaded Britain. Chapter 7 has discussed how these developments may have provided significant changes to the Time-Space constraints acting on Britain. Gaul and Britain engaged with each other long before the Roman period and the influence of Roman Gaul is generally seen as a catalyst for the political and economical developments in Britain during this time (Millet 1990: chapter 2). According to the conceptual model of innovation these conditions provided fertile ground for changes in beliefs (cause of innovation). This chapter will investigate which behavioural differences (effect of innovation) may be responsible for the variation between the outlier Iron Age assemblages. The results will be used to examine innovation during the process of Romanization.

Time and space across scales

The analysis of this faunal assemblage is complicated by the fact that it contains information relating to different temporal and spatial scales. Discussions of the confounding problem and the exploratory data analysis (chapter 9) have demonstrated the impact of scale on the interpretation of archaeological data.

One has also to remember that the human impact on animal bones entering the archaeological record goes beyond relatively short-time manipulation because subsistence strategies and the handling of livestock always reflects long term habitus and cultural beliefs.

Each of the faunal assemblages considered here has been influenced by Iron Age subsistence strategies which are long-term dynamics and differ between regions. Subsistence strategies change assemblage composition.

Also artefacts made from bone could have been in use for a long time before entering the archaeological record and one stratum may contain bones from different time frames. It is likely that their history is disguised and they simply become attributed to the temporal context from which they were recovered.

NISP counts are thus always a low-dimensional representation of a multi-dimensional history. The nature of the archaeological record makes it very difficult to unpick the different scales involved in its creation; however their recognition can provide valuable insight into the origin of an assemblage.

12.1 BEHAVIOURAL DIFFERENCES BETWEEN THE OUTLIER SITES

None of the analyses have identified any of the large assemblages with high frequencies of pig bones, like Ower or Skeleton Green. In addition neither PCA has indicated that pig bones were relevant for the outlier formation. Hence differences between the three groups can best be explained in terms of sheep and cattle husbandry and pig remains are excluded from further consideration.

Figure 51 gives a schematic illustration of the results so far. Because all vector loadings are very low, neither the separation of sites nor that of the skeletal elements are explicit. Nonetheless all cow elements have clustered in the same quadrant as the group 1 sites, while most sheep elements have fallen within the group 2 assemblages, which were generally less affected by taphonomic processes and partial recovery than the other outlier assemblages.

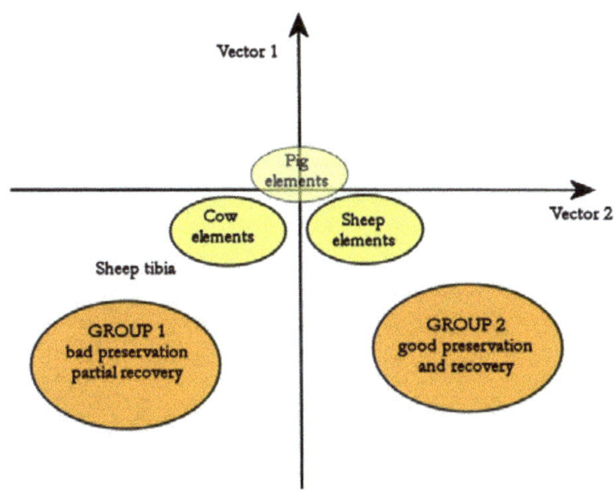

***Figure 51**: Simplified biplot of the outlier characteristics identified by two Principal Component Analyses*

Site characteristics

A first step will determine whether the site characteristics listed in Hamilton's original database (1999, appendix 2) can explain the separation of the outlier groups (table 20).

To get an impression of their geographical location, the 9 sites are marked on the map of Britain (figure 52). With the exception of Dragonby MIA-LIA which lies relatively far north and the Ditches hillfort in the South-West, all sites are distributed in the South and South-East of Britain. There seems to be no clear geographical division between the two groups. While the group 1 sites are spread throughout Wessex, the Upper Thames valley, East Anglia and the Midlands, both group 2 sites - Danebury and Maiden Castle - are located on the chalklands of Southern Britain. Five of the outliers are located in the chalklands of Wessex and are thus within the main trading area with Gaul during the first half of the 1st century BC.

The site elevation shows no distinct division between the groups. While both group 2 sites are hillforts and located between 76-150 m OD, the group 2 sites are mainly lowland settlements but also include the two hillforts, Ditches and Balksbury.

Figure 52: *Location of the 9 British Iron Age sites identified as outliers by PCA. •: group 1 sites ✶: group 2 sites. The small map shows all the sites in the original database (modified after Hambleton 1999: 15)*

There is a time factor to consider as well because not all the faunal remains date from the same period (table 20). Danebury, Rope Lake Hole and Cat's Water cover the entire Iron Age (~750 BC – AD 50), while the assemblages from Maiden Castle and Balksbury cover the Middle Iron Age (~ 400 BC – 100 BC) and the Dragonby remains are dated to about 100 BC to AD 50. The faunal assemblage included from the Owslebury site stretches from the 3rd century BC to the 1st century AD. The younger remains found at Puckeridge-Braughing and Ditches date from the Late Iron Age but a substantial part of the assemblages represents the second half of the first century AD.

Roughly speaking, the faunal remains of group 1 sites tend to be older than the group 2 assemblages, but the time scale does not allow a separation of the outliers.

Since none of the site characteristics divides the outlier groups, the following section will discuss which behavioural differences can be deduced from the faunal pattern itself.

Surplus, mode of production and hierarchical organisation

All the Iron Age sites that were emphasised by the statistical analyses have large faunal assemblages which suggest larger groups of people, possibly with a hierarchical organisation. The following section will relate the implications of this to subsistence strategies and the process of Romanization.

Group	SITE	Region	Geology	Height in OD	Date of sample	Site type
1	Balksbury MIA	Wessex	Chalk	76-150	MIA	Hillfort
1	Cat's Water	Eastern England & Anglia	Peat	0-25	IA	Open settlement
1	Ditches	Upper Thames Valley	Limestone	152-225	LIA-ERB	Hillfort
1	Dragonby	Midlands	Limestone	26-75	LIA	Open settlement
1	Owslebury	Wessex	Chalk	76-150	MIA-ERB	Open settlement
1	Puckeridge-Braughing	Eastern England & Anglia	Other	26-75	LIA-ERB	Open settlement
1	Rope Lake Hole	Wessex	Chalk	26-75	IA	Open settlement
2	Danebury	Wessex	Chalk	76-150	EIA-LIA	Hillfort
2	Maiden Castle	Wessex	Chalk	76-150	MIA	Hillfort

Table 20: *Site characteristics of the outlier sites (Hambleton 1999: appendix 2) OD: Topographical location of site in metres Ordnance Datum; EIA: Early Iron Age, MIA: Middle Iron Age, LIA: Late Iron Age, ERB: Early Romano-British, IA: Iron Age; Wessex includes Central Southern England*

Sahlins (1974: chapter 2) has made a connection between the mode of production and the social organisation of a group. He argues that groups that lack any significant social hierarchy do not fulfil their productive potential because "economics is only a part-time activity" (ibid. 86). Ethnographic data suggests that in non-hierarchical societies some available labour force remains unused, the division of labour is held to a minimum and available technology is not fully exploited. Sahlins classifies this type of economy as the *domestic mode of production.*

There is evidence for complete animal skeletons at Danebury which have been interpreted as animal sacrifices. Most of these burials are sheep and pigs, but horses and dogs were also found. While this might indicate surplus production, Grant (1984b) argues that horses were not as vital to the survival of the Danebury community as other domesticates, because - with the exception of speed - they did not offer any resources that other animals could not provide as well. Sacrificed animals were, in fact, individuals of little use to the community. In the case of sheep, pig and cow burials these animals are likely to have been the wrong sex for a particular exploitation strategy, or diseased. The evidence for a religious background for these animal remains is by no means inevitable. An Iron Age calf burial found in Dragonby has been interpreted as the possible trace of a children's pet: "It is tempting to imagine ritual or religious practices to account for the burial. There was no obvious evidence of ritual, however, either in the pit or around it" (May 1996: 126). It is possible that the complete skeletons found at Danebury simply represent natural losses that were buried in order to prevent smell. What is important here is the presence of complete animal skeletons. Such finds indicate that the population living at the sites did not face starvation, because it is reasonable to assume that old and ill specimens would have otherwise been killed before their condition made them inconsumable. The faunal evidence recovered from all 9 outlier assemblages suggests that these communities had advanced beyond the domestic mode of production and exploited the available resources beyond the level of primitive economies. This is supported by the evidence of long-distance trade with the continent (see chapter 7). Wine and other non-essential goods cannot be traded unless a significant surplus is produced that allows the exchange of some products.

Cunliffe (2000: chapter 15) has divided Iron Age Britain into different types of economy and argued that the 'hillfort-zone' from Wessex to the Welsh borderland produced a surplus and was engaged in intra-regional exchange. He has characterised this area as a *redistribution economy.* The economy of the southern coast probably shared the same characteristics but the subsistence strategies of settlements in Scotland did not produce a surplus. Cunliffe labels them as *suffic*er *economies.*

Of all the sites identified by the Principal Component Analysis the Dragonby settlement is the furthest north. It is located near Scunthorpe in Lincolnshire and was thus well beyond the main trading area with the continent at the time. Still its Iron Age layer produced Gallo-Belgic or related pottery, metalwork and some silver coins. These finds suggest its occupation by people with a higher social standing (May 1996: chapter 4 & 5). Faunal remains in Puckeridge-Braughing were recovered from an Iron Age settlement known to be a wealthy trading centre and the surrounding area is considered to have hosted a substantial woollen industry (Fifield 1988). Ditches hillfort is a high-status site with a very early villa (Trow 1988). The site at Rope Lake Hole was used for salt (briquetage) and shale industries. It lies in close proximity to Eldon's Seat which had to be excluded from the statistical analysis due to its incomplete representation of skeletal elements. Although the species distribution is similar for both sites, other artefacts have suggested that they had different 'economies'. Rope Lake Hole had a more 'industrial' character while the activities at Eldon Seat focussed on domestic production. (Woodward 1987: 145-149).

These examples show that many of the outlier sites show evidence of an advanced economy with signs of surplus production or specialisation and groups organised in a hierarchical way. Both factors have been identified as favourable for successful Romanization linking the statistical analysis with the literature review. Nevertheless because these characteristics are found throughout all assemblages they cannot to explain the division of the outlier groups.

The identification of large sites could also be a matter of sample size rather than selection for internal hierarchal organization. Balksbury Camp was represented by two faunal assemblages, a small early Iron Age sample with 191 bone fragments and a larger one with 3037 bones dating from the Middle Iron Age. Only the latter was identified as an outlier, which is probably due to its high frequency counts. It is unlikely that the temporally later and larger sample indicates internal developments over time which resulted in more complexity, because there is no settlement evidence for Balksbury Camp during that period (Wainwright & Davies 1995: 19).

The age distribution of the kill-off pattern

With the exception of Rope Lake Hope all, outlier site reports contain information about the age of the main domesticates represented in the assemblage (see table 16, p. 83). These mortality profiles might provide some indication about the subsistence strategies applied at the sites.

In his comparison of the Neolithic Revolution in the New and Old World, Sherratt (1981) concluded that the main difference between both regions was that people in the old world also exploited domesticated animals for milk, wool and traction rather than just meat. He argues that such a broad exploitation of animal resources provided a surplus in energy and allowed for the subsistence of much larger groups; milk, for instance, provides 4 to 5 times as much protein and calories as the consumption of the animal's meat with the same fodder input. The secondary use of domestic animals also had an impact on agriculture, transport and mobility. Plough use, for instance, increased grain production and allowed the exploitation of less fertile soils. Although urbanisation is possible without a 'secondary product revolution', Sherratt claims that the energy surplus provided by it clearly accelerated developments in the Old World: "Cultivation alone, without the extensive use of domestic animals, was able to sustain even complex urban societies. But it is not without significance that the next threshold, that of industrialisation, was attained only in the Old World" (Sherratt 1981: 261). Like Sahlins, Sherratt links the mode of production to the social organisation of a group by saying that "anthropologists have long recognised the importance of plough agriculture as a predictor of social systems involving new mechanisms of inheritance" (ibid: 263). The link between social hierarchy and economic development has also been made for Gaul: "The Gallic economy might have been different, but it was far from being primitive. The rich grave-goods of the Hallstatt and La Tène burials demonstrate that a surplus was being produced which was being used to develop a sophisticated specialisation of skills and roles [...] and so a social hierarchy [...]" (Drinkwater 1983: 10).

The use of animals for secondary products rather than meat is likely to result in different kill-off strategies. Milk production depends on lactating and thus older female animals which will consequently appear more frequently in the archaeological record of mixed ecnomies than in the remains of purely meat producing economies. While animals that are used for meat production are expected to be killed relatively young to maximise the yield, animals that are used for traction, milk or wool are killed off later in life. Thus the age distribution of a faunal assemblage might give evidence about the ways in which these animals were exploited. For sheep the optimum age for meat production is 1½ to 2½ years, for cattle it is about 2 years. Castrated males have been used for traction while the production of milk requires more females. The gender distribution can therefore also give evidence about husbandry practices (Sherratt 1981: 283-285, Grant 1984a, Maltby 1994). Mortality profiles are generally determined by epiphyseal fusion and/or teeth eruption and, after all teeth have erupted from teeth wear.

In Maiden Castle over 50% of the sheep population survived to at least the age of 5, which indicates their exploitation for milk and wool. The sex of 17 sheep could be determined and the majority of them were female, which supports this assumption (Armour-Chelu 1991). Sheep in Balksbury were either killed at a young age or when mature (Maltby 1995) and it is likely that the second group was also used for milk and wool. This is supported by the presence of circular chalk loom weights at the site (Wainwright & Davies 1995: 19). Sheep remains in Puckeridge-Braughing show two similar peaks in the mortality pattern, one after the third year and another at a comparatively young age. The latter has been interpreted as the use of male animals for the production of tender meat but could also represent natural losses. The older animals are thought to have been used for wool production (Fifield 1988). About a third of the sheep bones found in Danebury are from mature animals and thus beyond the optimum age of meat production. Again this pattern, together with the loom weights, spindle whorls and weaving combs discovered at the site, has been interpreted as wool production (Grant 1984a). The presence of neonates and burials of young animals at Danebury, Maiden Castle and Balksbury suggests that calving and lambing took place at the sites where they would have been protected from predators like wolves (Grant 1984a, Maltby 1995, Armour-Chelu 1991). Generally these four sites show a kill-off pattern in sheep that reflects the natural loss of very young lambs and the use of older animals for wool and milk. It is likely that the sheep were kept at the site for at least some time during the year.

The other sites show a slightly different pattern. At Dragonby the sheep assemblage shows several mortality peaks, one at 6 months, a second at 18 months and a third at 30 months of age. The second peak is likely to represent animals slaughtered for meat. The number of very young animals found at the site is comparatively small and lambs especially appear to be underrepresented. It has been suggested that sheep were kept outside the settlement and animals that died shortly after birth were left behind. Pigs and cattle on the other hand were kept near the settlement, or at least calving and farrowing took place inside the settlement, which would explain the presence of small calves and piglets in Dragonby (Harman 1996). The sheep distribution at Ditches hillfort shows two mortality peaks; about 40% of the animals were killed at 1 to 2 years and about 50% at 2 to 3 years (Rielly 1988). The first peak coincides with the optimum mortality for meat production and the low number of lambs below 1 year of age might indicate that the animals were not kept at the hillfort. Because of the high status associated with the site, it has been suggested that the sheep remains might represent bought-in food (Hambleton 1999: 73). One third of the sheep recovered at Rope Lake Hole died between 1 and 17 months of age, one third between 18 and 31 months and another third after the age of 32 months (Coy 1987). This also implies the use of sheep for meat while the mature animals might have been exploited for secondary products. The sheep assemblages at Owslebury (Maltby 1987) and Cat's Water (Biddick 1984) also show a greater proportion of sheep killed at an age most suitable for culling for meat.

Thus in Dragonby, Rope Lake Hole, Owslebury, Cat's water and Ditches the emphasis appears to have been on sheep meat rather than its secondary products. The animals are also likely to have been kept outside the settlement which differs from the strategies practised at Maiden Castle, Danebury, Puckeridge-Braughing and Balksbury. While both group 2 sites show great similarities in the age distribution of sheep, the husbandry strategies do not allow a separation of the two outlier groups, because some of the group 1 sites also fit into the same pattern. Sheep husbandry is thus not likely to be responsible for the statistical identified differences between the groups.

The kill-off pattern for cattle shows less variety than for sheep and is very similar for all sites. For the cattle recovered at Rope Lake Hole no mortality profile could be determined (Coy 1987). About half of the cattle bones from Owslebury that allowed an age determination were recovered from the 3rd century ditches and contained a great proportion of immature animals, which indicates the exploitation of meat. The majority of the remaining finds were over 40 months old which indicates the use for traction (Maltby 1987). All other sites had the majority of cattle killed at a mature age which indicates that this species was mainly used for traction and manure rather than for meat (Grant 1984a, Armour-Chelu 1991, Harman 1996, Maltby 1995, Rielly 1988, Biddick 1984). It is generally accepted that in prehistoric societies plant food was a more important source of calories than meat and agriculture was certainly crucial throughout the Iron Age (Grant 1984a). The mortality profile of the outlier sites supports this assumption. Cattle do not appear to be reared especially for meat and manure will have played an important role in keeping the land fertile for the cultivation of grain. Danebury shows high frequencies of first year cattle mortality. While Iron Age cattle are not thought to have given milk beyond the time the calves were weaned, only cows that lost their calves would have provided milk for humans (Grant 1984a). The high frequencies of calf losses might indicate a deliberate killing in order to increase the milk yield available for humans. Harman (1996) has suggested the same explanation for the mortality peak for calves found in Dragonby. These remains of very young animals are likely to at least partly represent natural deaths as well. The Cat's Water assemblage only contains a small population of calves and it has been suggested that calving off site could be responsible for this pattern (Biddick 1984).

The age distribution of cattle and sheep remains does not allow a separation of the two outlier groups based on differential husbandry regimes. Both species appear to have provided secondary products and especially older cattle supported a relatively high-level agriculture. There appear to be differences in sheep husbandry as about half of the outlier sites used them for the production of meat and the other half put a stronger emphasis on the exploitation of secondary products (see table 16, p 83). This separation is independent from the grouping created by the Principal Component Analysis, but none of the group 2 sites seems to have used sheep primarily for meat production. Thus sheep were exploited for wool, and at some group 1 sites, also for meat while cattle were used mainly for milk and traction.

Recovery context

Differences in the distribution of skeletal element that were not caused by post-mortem processes may give an indication of the use of a particular site. If butchering was the main activity at a site, low meat-bearing elements are likely to appear in greater frequencies while the high meat-bearing elements are expected to be more common at consumption sites. Production for trade would lead to an under-representation of the exported species or its meat-bearing elements. Of course, settlements as large as most of the outlier sites are likely to show intra-site variation and have different areas allocated to different activities. This information is lost at the level of inter-site comparison used for this study. Most of the faunal remains of the outlier sites were recovered from pits and ditches but, with the exception of Danebury, Owslebury and Rope Lake Hole, none of the site reports gives a detailed discussion of the deposit type in which the animal bones were recovered which makes a more detailed discussion of the different deposit contexts impossible.

Differential preservation has severely limited intra-site analysis of disposal patterns in Owslebury. An analysis of the less affected Iron Age layers did show differential preservation between deposit types; sheep bones were better preserved in pits while cattle were more common in the ditches. This pattern has been interpreted as evidence for differential butchery practices: larger mammals were slaughtered away from the main living areas and their remains deposited in the marginal ditches (Maltby 1987). The mortality profile for sheep varies between pits and ditches. Young lambs are rarer in the ditches, which might be the result of human behaviour or the fact that bones are generally less well preserved in ditches which reduces the survival rate of the fragile bones of younger animals.

Ranking of skeletal elements

Next the rank order of skeletal elements is analysed in regard to inter-group differences. Once the mandibles and phalanges are left out, the overall rankings of the skeletal elements show great similarities between the two groups (see figure 53 & 54). With the exception of the sheep tibia in group 1 the accumulated variables x_foot and x_radul are the most and second most frequent cow and sheep elements for both groups. The cow tibia is also slightly less common in group 2. The sheep pelvis ranks last in group 1 and third last in group 2, but all other elements differ by no more than one rank. Thus only the sheep tibia remains as a differential feature of the two groups.

The more detailed rank order for the individual sites shows the same similarities. Generally the sheep femur is more common in the group 1 than in the group 2 sites while the sheep scapula is more frequent at the group 1 sites. This does not suggest that the carcasses were treated differently at the sites.

An inter-species comparison shows that for sheep, with the exception of the femur, the limb bones are more abundant than the axial skeleton. This overall pattern appears in both outlier groups but is not visible in the distribution of cow elements. It suggests that the sheep assemblages have experienced different treatment from the cattle. The culling for meat was more common in the sheep assemblages (see table 16, p 83) and might be responsible for the higher presentation of sheep limb bones.

This result is consistent with the site reports; the representation of the main domesticates suggests that none of the outlier sites experienced a significant separation of the carcasses. Rope Lake Hole shows balanced frequencies for both meat- and non-meat-bearing bones (Coy 1987) and in Puckeridge-Braughing every cow element is represented (Filfield 1988). Analyses for Maiden Castle and Danebury also suggest that the bones recovered at the sites come from whole animals (Grant 1984a, Armour-Chelu 1991). Such a complete representation of the skeletons suggests that slaughtering took place at the sites, making the by-products of this activity (for instance horns or hide) available for secondary use. It is thus unlikely that meat was exported in significant quantities. The only exception is Ditches; while none of the skeletal elements is significantly over- or under-represented at the site, its under-representation of small lambs has been interpreted as evidence for sheep importation (Rielly 1988).

Species distribution and Romanization

Mode of production, ranking of skeletal elements and mortality profiles have produced no criteria that can be used to differentiate the two outlier groups.

However, the statistical analyses have already indicated that the species distribution might allow a clearer separation (see table 6, figure 38). Sheep is the most abundant species in 7 of the 9 outlier sites. This dominance might be explained by geography, considering that the majority of sites are located on the Wessex chalk lands, an area that is traditionally associated with sheep farming because its grass is only of medium quality which makes it less suitable for cows (Maltby 1994, Grant 1984b). For Danebury about 60% of all the identified bones were sheep bones (Grant 1984a) and for Maiden Castle the proportion is even higher with 66% (Amour-Chelu 1991). Balksbury, Owslebury and Rope Lake Hope, the group 1 sites that also lay within the chalk lands are also dominated by sheep. However this pattern cannot merely be explained in terms of environment. Firstly two of the group 1 sites, Puckeridge-Braughing and Dragonby, are on different soils but still dominated by sheep and secondly it is in fact the *proportional* distribution rather than their species rank order which separates the groups: sheep is proportionally less common in the group 1 assemblages, including that group's chalkland settlements. Table 21 shows that the average frequency for sheep is considerably higher in group 2, while cattle are proportionally more frequent in group 1.

This reflects a behavioural difference between both groups: sheep were kept in relatively higher quantities in group 2 than in group 1. The husbandry regime of the British Iron Age has generally been associated with high sheep frequencies (see section 8.3): even though there are regional and intra-site differences, sheep were kept most frequently although cattle provided most meat for the diet (Ryder 1983: 74-83, Maltby 1994, Hambleton 1999: chapter 6). Therefore the species distribution at the group 2 sites, Danebury and Maiden Castle, both important Iron Age hillforts, can be considered as typical for that time. Both assemblages include early Iron Age contexts and are thus generally younger than the group 2 assemblages.

But even the later phases of Danebury still contain proportionally more sheep than the contemporary group 2 sites (see table 21). The Ditches hillfort deviates from this generally observed pattern of Iron Age husbandry because of its proportionally high numbers of cattle bones for the Upper Thames area. This has been interpreted by Rielly & Trow (1988) either as an indicator for an incipient 'Romanization' of the diet or as an exceptionally fast 'Romanization' within only a decade of the conquest.

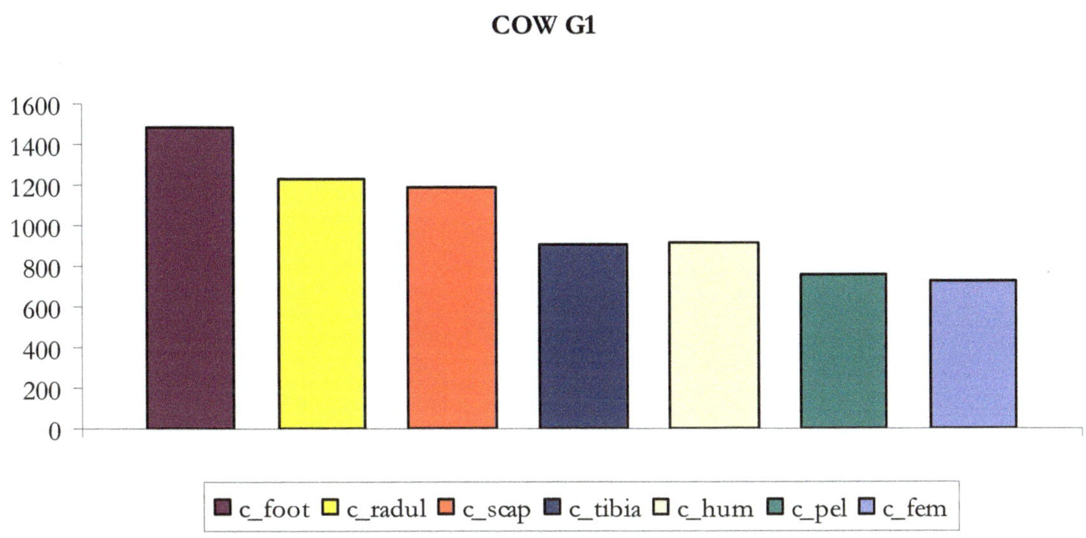

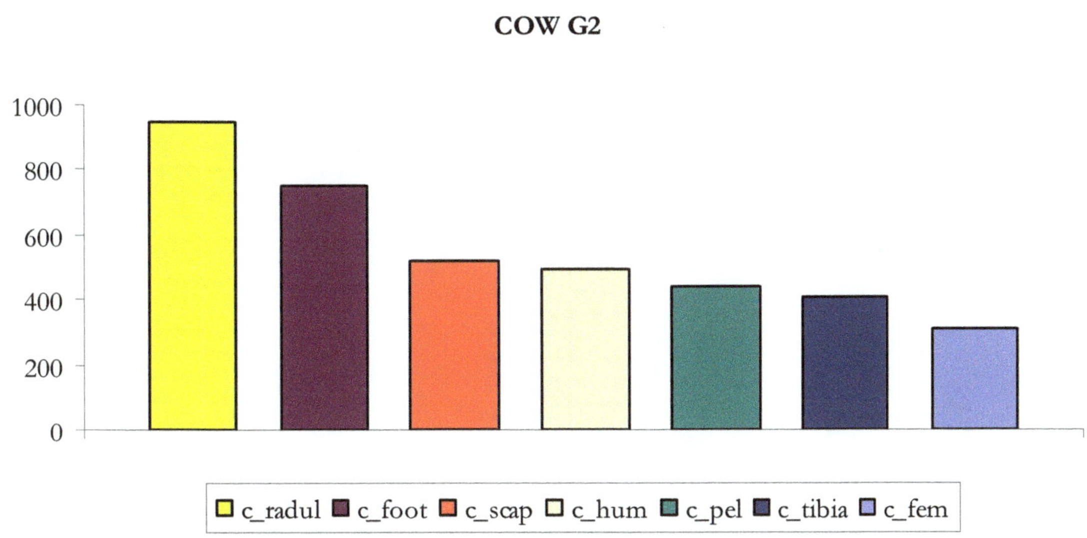

***Figure 53**: The rank order of the skeletal cow elements without mandible and phalanges*

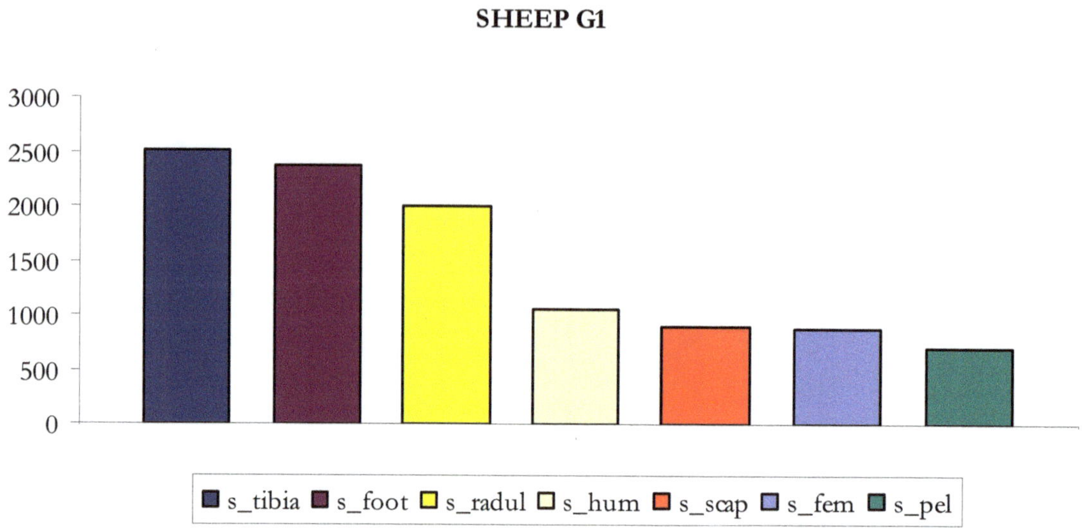

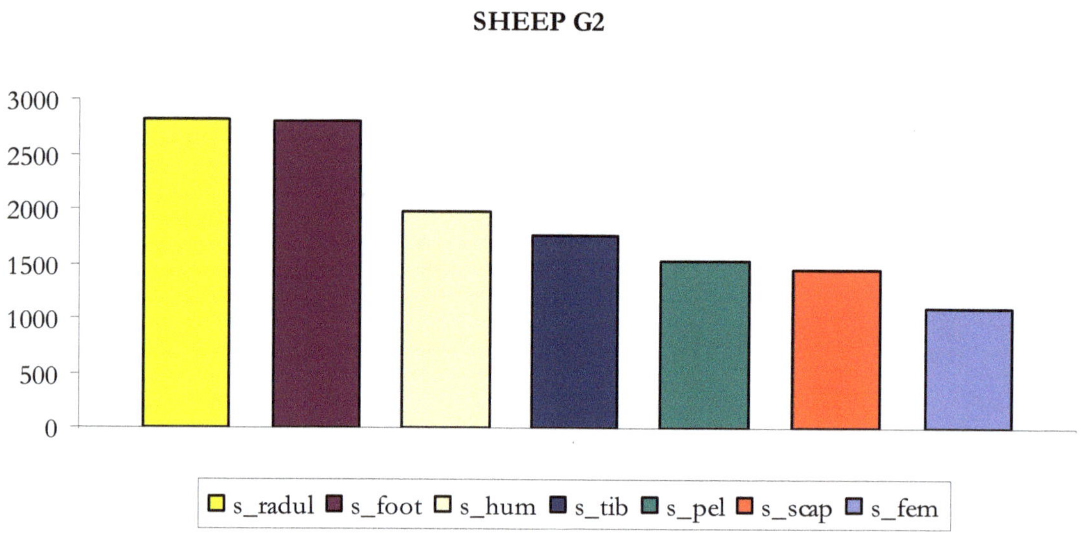

***Figure 54**: The rank order of the skeletal sheep elements without mandible and phalanges*

SITE	Sheep %	Cow %	Pig %	Sheep $_{COR}$ %	Cow $_{COR}$ %	Pig $_{COR}$ %	Reference
Group 2							
Danebury EIA	56.0	23.0	15.0	59.6	24.5	16.0	Grant 1984a
Danebury MIA	57.0	16.0	18.0	62.6	17.6	19.8	Grant 1984a
Danebury MIA-LIA	58.0	16.0	14.0	65.9	18.2	15.9	Grant 1984a
Danebury LIA	66.0	19.0	9.0	70.2	20.2	9.6	Grant 1984a
Maiden Castle	66.0	21.0	9.0	68.8	21.9	9.4	Armour -Chelu 1991
Group 2 mean and s.d	60.6 ± 5	19 ± 3.1	13 ± 3.9	65.4 ± 4.4	20.5 ± 2.8	14.1 ± 4.5	
Group 1							
Puckeridge-Braughing	39.3	31.1	24.0	41.6	32.9	25.5	Fifield 1988
Balksbury	43.1	24.7	31.1	53.1	30.4	16.5	Maltby 1995
Dragonby	49.8	27.5	13.0	55.2	30.5	14.4	Harman 1996
Owslebury	44.7	32	15.7	48.4	34.7	16.9	Maltby 1987
Cat's Water	42.7	49.8	7.5	42.7	49.8	7.5	Hambleton 1990
Rope Lake Hole	56.0	37.0	7.0	56.0	37.0	7.0	Coy 1987
Ditches	38.0	47.0	15.0	38.0	47.0	15.0	Rielly 1988
Group 1 mean and s.d	45.34 ± 24.8	35.69 ± 9.6	14.88 ± 9.8	47.85 ± 7.2	37.47 ± 7.9	14.7 ± 6.3	

Table 21: *Species distribution of the three main domesticates for the outlier sites. Because of other species found at most sites the frequencies of the three main domesticates do not add up to 100%. In order to directly compare the frequencies they are corrected and x_% $_{COR}$ is the species % multiplied by f, whereby: cow % + sheep % + pig % = x and 100/x = f. Sheep $_{COR}$ + cow $_{COR}$ + pig $_{COR}$ = 100%*

The majority of its cattle were at least 4 years old at the time of death (Rielly 1988) which suggests that they were used for traction or dairy products. This suggests substantial agriculture at Ditches hillfort, supported by cattle-drawn ploughs and fields fertilized with manure.

This result links with another distinguishing factor between the groups, one that becomes apparent on a different time scale. While all the sites in group 1 were occupied beyond the Iron Age and have produced a number of Roman artefacts in the later layers, both group 2 sites show no signs of Romanization. The occupation of Danebury as a strongly defended settlement appears to come to a sudden end around 100 BC (Cunliffe 1984: 550). This coincides with the disruption of various other hillforts and the development of the first oppida in South-East Britain (see chapter 7.6). The internal organization of Maiden Castle was also abandoned around the same time and settlements appeared close to the hillfort (Sharples 1991b: chapter 9).

Species distribution has been linked to Romanization before. King's (1984, 1999) comparative studies of species distribution for the late Iron Age and early Roman period in Britain have shown differences between Romanized and non-Romanized sites. While the average sheep proportion for high-status sites in South-East Britain is 31.3 % ± 4.7, the value for non-Romanized sites is higher with an average of 47.8% ± 13.7. Thus there is evidence that the species distribution changes after the Romanization of Britain. King (1984, 1999) has argued that the British Iron Age subsistence pattern adapted to the probably already established diet of the Roman army which contained higher proportions of cattle and pig. This observable shift in species distribution correlates with other factors seen as indicators for Romanization at these sites (table 22). King interprets the pattern as follows "This appears to show that the urban, military, and legionary sites set a dietary pattern (presumably derived from Gaul and Germany) that was emulated by social groups seeking to become more Romanized" (ibid: 180). Thus he argues that higher cattle frequencies reflect a conscious adaptation to Roman demands by British communities.

The results represented here suggest a different picture, since the PCA has identified differences in species proportion for *Pre-Roman* assemblage; because the analyses were based exclusively on Iron Age data the identified pattern cannot have been caused by a Roman presence in Britain. All the group 1 sites contain proportionally fewer sheep than the group 2 sites. This suggests that the Roman presence rewarded the existing Iron Age subsistence strategies differentially by favouring a more cattle-based economy but did not initiate a stronger emphasis on cattle. Even though the presence of the Roman army might have accelerated the development towards a more cattle-based economy, this process windowed variation in subsistence strategies already existing during the Iron Age.

The patterns of the faunal remains represent the forces of supply and demand rather than a copying of Roman lifestyles.

CATEGORY	SAMPLE SIZE	CATTLE	SHEEP/GOAT	PIG
Late Iron Age	10	39.0 ±14.9	31.3 ± 8.5	29.7 ± 14.5
Roman				
Towns	50	53.5 ± 18.5	27.0 ± 14.0	19.5 ± 9.8
Vici	69	56.3 ± 17.6	31.9 ± 15.5	11.8 ± 7.7
Villas	58	55.6 ± 15.8	29.8 ± 12.2	14.6 ± 11.7
Rural	90	47.1 ± 15.7	41.2 ± 15.7	11.7 ± 9.0
Legionary sites	16	63.5 ± 16.8	14.9 ± 11.2	21.6 ± 11.0
Auxiliary sites	63	64.8 ± 19.7	22.3 ± 17.2	13.0 ± 8.4

Table 22: *Species distribution of Romano-British sites in percentage mean ± standard deviation (King 1999: 180)*

The change towards a more cattle-based economy was by no means compulsory (sheep farming remained viable) and the overall pattern for Roman Britain (King 1984, 1999) disguises individual variation by collapsing sites. The faunal remains for Rope Lake Hole show a decrease in the proportion of cattle from 37 to 23% while sheep increase from 56% in the Late Iron Age to 71% in the Roman period (2nd and 3rd centuries AD) (Coy 1987). However there is an overall trend towards a stronger emphasis on cattle. Owslebury shows evidence that sheep became less important during the Roman period (Maltby 1987) and the proportion of cattle increases from 19.4 % in the Late Iron Age to 35.6 % in the late Roman contexts (Maltby 1995).

The statistical analysis based on the conceptual model of innovation has separated the largest assemblages in the database into settlements that declined at the end of the Iron Age and settlements that remained occupied and eventually became Romanized.

12.2 CONNECTION TO THE CONCEPTUAL MODEL: The Romanization of Britain as a complex multileveled process placed in the physical landscape

The exploratory data analysis and the follow up statistics of the observed pattern have produced evidence for differences in behaviour regarding Iron Age subsistence strategies. We must now relate these patterns to the conceptual model of innovation on which this case study was based. The first part of this book suggested a number of hypotheses regarding the process of innovation. These will be rejected or verified for the Romanization of the North-Western Provinces based on the results discussed above.

It should be remembered that the time lag between the invisible change of beliefs and the observable change in behaviour means that the actual process of innovation is over by the time we can observe its effects (Trott 2002), especially in the archaeological record. We also have to consider the aspect of temporal scale in the connection of both analytical parts. The faunal remains were recovered from different temporal contexts ranging from 750 BC (Early Iron Age) to the second half of the 1st century AD (early Roman Period), while the literature review on the Romanization of Gaul begins at 120 BC and thus covers a shorter time-period than the empirical data. This fact does not allow us to make any assumption about the origin of the differences in the observed pattern. The remains from the group 2 sites range from about 750 BC to 100 BC while the group 1 site assemblages are slightly older and cover the time from 400 BC to AD 100.

The next section will discuss the developments in Britain with regard to the Roman progression in the North-Western provinces, considering patterns *before* the Romans moved into the area, *during* their campaigns in Gaul and *after* they occupied Britain.

Pre-Roman pattern (500 BC to 120 BC)

One of the hypotheses made about the Romanization of Britain is *that key changes took place before the Romans occupied the island.* All the outlier sites are large settlements and thus fulfil important prerequisites for an 'easy Romanization' (agricultural intensification, high population density, social stratification). Their size, combined with evidence of other artefacts, indicates that the economy and social organization of the groups living there must have reached a level of complexity which made them more accessible for the Romans. La Tène style metalwork developed in Europe around 500 BC, was found in Britain and indicates the trade of prestige objects by Iron Age élites. This must have been facilitated through surplus production at some of the British Iron Age sites. Cunliffe (2000: chapter 15) has described the Wessex zone, where the majority of outlier sites are located, as a *redistribution economy*, based on its engagement in intra-regional exchange. This mode of exchange demands the existence of some form of coherent political organization, a chief or big man. The redistributive agents in a chiefdom are responsible for the regulation of trade and the distribution of certain resources. This form of exchange enables individual communities to overcome the limitations of their own subsistence strategies by trading for the products they do not produce themselves (Sahlins 1974: chapter 5, Renfrew & Bahn 1991: 307-314). The archaeological evidence points towards well-established trade between Britain and the continent for the 5th to the 2nd century BC (Cunliffe 2000: chapter 16).

This connection between the mode of production (surplus production and redistribution) and social hierarchy (chiefdom) fits into Sahlins' (1974) argumentation about economical development discussed in section 13.1.2. It also links subsistence strategies manifest on the local level to processes of inter-regional trade which can only be observed on a regional scale or higher.

The empirical analysis was based on faunal remains of Iron Age Britain and thus identified Pre-Roman patterns. The fact that the Principal Component Analysis separated all the sites that later became Romanized from the sites that did not, suggests that the variety in subsistence strategies that existed before the Romans moved into Britain may have become significant later. These different patterns might reflect the effects of previous innovations and the question about their origin should not be neglected. The PCA has separated two different subsistence strategies for large, high-status sites in the British Iron Age; some sites had proportionally more sheep (group 2) which were less abundant in others (group 1). Cross-channel comparison has shown that the subsistence strategies in Iron Age assemblages of Northern France differ from the pattern in Britain in as much as pig is the most abundant species (Hambleton 1999: 44) (see figure 55). We can therefore conclude that subsistence strategies were not brought into line through Pre-Roman contacts between these two Iron Age communities and that in this case the exposure to different means of making a living did not directly lead to a change of behaviour. In fact these differences in subsistence strategies between high-status sites might even have been part of the basis for the cross-channel trade patterns during the Iron Age. The pork produced in Gaul could have been exchanged in Britain for wool produced at group 2 sites with proportionally more sheep and for grain produced by group 1 sites with cattle used to support mixed agriculture.

Pre-Roman trade between Iron Age Gaul and Britain is likely to have been constrained by negative feedback loops that limited demand on both sides. Goods that represent prestige objects cannot be traded in great quantity otherwise they lose their special nature. Additionally, existing Time-Space constraints meant that low-status subsistence goods could not be transported effectively in great quantities beyond the incipient area, because the recipient market would be saturated quickly.

Transition period: the Romans settle in Gaul (120 BC to AD 43)

The archaeological evidence suggests that trade between Britain and Gaul expanded significantly during the 1st century BC. This intensification has been linked to the activities of Roman traders in Gaul and various other developments connected to the Roman presence in this area (Cunliffe 2000: chapter 16, Millett 1990a: 29-33, Drinkwater 1983: 38). This traditional view will be discussed with consideration of the new insights gained from this case study.

As Drinkwater pointed out, the Gallic Iron Age people were perfectly capable of using roads and waterways for trade but the Romans, by "establishing a planned, measured, mapped, advertised, maintained and policed all-weather network, and in providing that comprehensive political and administrative framework [...] gave the Three Gauls the integrated system which they previously lacked" (1983: 124). The Romans did not establish long-distance trade in Gaul but Romanization altered the nature and pattern of existing trade by facilitating long-distance transport and making the recipient market harder to saturate. Therefore the modifications in Gaul did not provide new but rather *quantitatively different,* contacts to Britain.

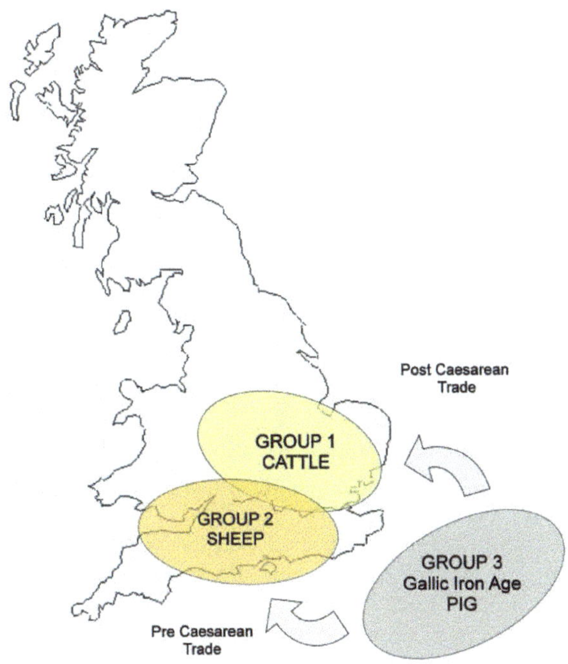

Figure 55: *Schematic representation of the variety in subsistence strategies of high-status sites in the British and Gallic Iron Age. There is no geographical separation of the subsistence strategies in Britain*

Britain was evidently not a closed society nor isolated from the continent and thus was influenced by the extension of roads, trade and administration in Gaul, which constitute changes in Time-Space constraints. It is reasonable to assume that people living in Iron Age Britain experienced some of these impacts as changes in their environment. The assumption that *innovations are more likely to occur in areas that are influenced by changes in TS constraints than static environments* makes Britain a likely candidate for social change.

Sheep husbandry in Britain did not disappear during this transition phase but the high-status sites in this study that survived into the Roman period were the ones with a proportionally higher share in cattle. When Roman traders moved into Gaul they created a market and demand. This was crucial for British communities producing a surplus, because it increased their opportunities for trade with different objects and different partners. As trade between Britain and Roman Gaul amplified, it created positive feedback loops because the increased sustainability of trading opportunities would have provided an incentive for both sides: for Rome to change the infrastructure in Gaul in order to facilitate trade with Britain and exploit its resources and for British communities to adapt to Roman demands. During this transition phase sheep keepers in Britain (group 2 sites) ceased to be potential high-status clients for traders in Roman Gaul which triggered the changes observed in the post-Roman pattern. The chiefdoms of the British Iron Age either fitted into the Roman market created in Gaul or they disappeared.

The variety in local subsistence patterns appears to decrease during the process of the Romanization of the North-Western provinces when certain ways of making a living (a stronger emphasis on sheep husbandry) became unattractive. This is unlikely to have been solely caused by environmental factors or coercion by the Roman army. It is reasonable to assume that subsistence dependent on a higher proportion of sheep would have still been sustainable under Roman rule and the change of subsistence pattern may represent choice or a change of beliefs. The presence of the Romans is likely to have provided incentives that made certain subsistence strategies more attractive as they were better suited to fulfil Roman demands and thus allow for trade with them. The changes would have been brought about by changes in the Time-Space constraints (in this case, Romans creating opportunities for trade) and the responses to these changes would have been partly elective. It is possible that these changes led to a reorganization of the landscape, because the requirements of cattle are not the same as those of sheep.

It has been argued that, on the imperial level, the expansion of the Roman Empire was motivated by a quest for prestige created by a highly competitive system of power in Rome rather than economic incentives. After their first campaigns the Romans did not show any attempts to organize a systematic exploitation of the new territories. However this appears to have changed during the Augustinian period (63 BC - AD 14) which saw a reorganization of the Gallic provinces. The Gallic road system that connected the inland areas with the coast was constructed around the same time (40 BC). Augustus also established formal treaties with some British tribes (Millett 1990a: 29-35 & chapter 10). These developments will have relaxed the Time-Space constraints between Gaul and Britain, whose growing markets of supply and demand stimulated each other in positive feedback loops and thus created a multiplier effect that stimulated an innovation cascade. The analysis of faunal remains shows one way of reacting to these changes was a shift in subsistence strategies, from sheep-based to mixed farming with a proportional increase in cattle production.

Post-Roman Conquest Pattern (after 43 AD)

After the establishment of more substantial trade between the two countries, the Romans invaded Britain in AD 43[1]. This provided an opportunity to further relax the Time-Space constraints acting on cross-channel trade through improvements of the British infrastructure. Claudius' campaigns in Britain are thought to have brought material wealth. They increased the opportunities for trade as an effect of the relaxation of T-S constraints.

British Iron Age communities are considered to be fundamentally agrarian and grain remained the basis of the British economy throughout the Roman period (Cunliffe 2000: chapter 15). The shift from sheep to cattle husbandry that has been observed in Roman Britain could represent stronger emphasis on agriculture as cattle have been used both for traction and the manure for fertilizing the field. Grain and cereals would have provided goods to exchange with the Roman Empire.

We have explained the hypothesis that innovation *is a complex process which includes top-down constraints as well as bottom-up behaviour*. Not all the South Wessex sites, characterised by geographical similarities, a close proximity to Gaul and recorded trade contacts with the continent throughout most of the Iron Age, showed the same reaction to changes in this area. While Rope Lake Hole and Balksbury Camp continued to be settlements and eventually became Romanized, occupation at Danebury and Maiden Castle declined. The comparison of species proportions has shown that they belong to different subsistence groups and while both forms existed alongside each other throughout the Middle Iron Age, husbandry regimes with a stronger emphasis on sheep declined after 100 BC.

The settlement pattern shows a different response to Romanization from the subsistence strategies; 4 of the 9 outlier sites are classified as hillforts. The classification in Hambleton's case study seems to be based on the geographical location of this settlement type because, unlike the oppida, the hillforts are all placed at a higher altitude of at least 76 m OD. Two of these hillforts (Balksbury and Ditches) are clustered in group 1 and became Romanized while the other two (Maiden Castle, Danebury) constitute the second outlier group and show no traces of full-scale Romanization. The settlement evidence in Maiden Castle shifts and small undefended homesteads appear in the area around the hillfort while its occupation gradually breaks down (Sharples 1991a: 257-265). Occupation of Danebury declined around 100 BC (Cunliffe 1984: 550). Whatever factors made these two sites unattractive after 100 BC did not affect Balksbury and Ditches which remained hillfort settlements. The

[1] Again political prestige rather than economic incentives predominate as the reason for Claudius invasion of Britain in the Roman literature. However the effects of his campaigns on the British economy are nevertheless the same.

observation scale of this study does not allow more detailed assumptions about the causes behind these changes in settlement pattern.

The first part of this book argued that major social change (and changes that manifest themselves in the archaeological record are mainly large scale) is often connected to a cascade of innovation. The processes described above support this idea; there was not one process of Romanization but several interconnected changes and the *Romanization of Britain was not caused by a single innovation but rather by a cascade of change and positive feedback loops* on different temporal and spatial scales. This is closely connected to another characteristic of innovation; *it is a complex process whose effects might show at different levels and over different time scales.*

Could the case study reveal innovation?

This case study was aimed at testing the conceptual model developed in Part 1. This required evidence for changes in Time-Space Constraints indicating changes in beliefs as well as evidence for behavioural changes.

The data did not allow a chronological separation of Early to Late Iron Age assemblages. Therefore differences in animal husbandry strategies do not represent changes in behaviour as such, but are likely to represent the results of earlier innovations. However a shift in observation scale through the use of site reports revealed changes in behaviour over time in as much as the sites with a stronger emphasis on cattle became Romanized. Various tests established - as thorough as the archaeological data allows - that the pattern discovered within the faunal assemblage is not merely the result of taphonomic processes or partial excavation but actually represents the impacts of human behaviour (see chapter 12).

Recall that the relationship between changes in Time-Space Constraints and changes in beliefs is one of probabilities (see section 7.1). The results of this study can therefore not *prove* that innovations took place during the Romanization of the North-Western Provinces. However it has been shown that there were changes in Time-Space Constraints (for instance the impact of the developments in Gaul) followed by changes in behaviour (high status sites with a stronger emphasis on sheep husbandry discontinue) which according to the conceptual model make it *very likely* that innovations were behind these processes.

This case study was aimed at testing the conceptual model developed in Part 1. This intention required evidence for changes in Time-Space Constraints indicating changes in beliefs as well as evidence for behavioural changes.

Previous concepts of Romanization in regard to the research results

Even though this analysis did not intend to gain insight into the identity and belief systems of indigenous people during the Roman transition, it can make an important contribution to the debate on the Romanization of Britain. It has become very clear that the process of Romanization took place on various different temporal and spatial scales which is in accordance with the claim that "Romanization has many different forms and encompasses all of the consequences [...] brought about by the infiltration of the Roman world" (Meadows 1994: 133).

The changes in Time-Space Constraints imposed different effects on the indigenous people of Britain depending on their position in society. It seems that the process of Romanization was by no means one-sided or simply forced upon the indigenous population. Similar changes in top-down constraints can trigger or allow for different bottom-up reactions. It also shows that there is never only one constraint acting on a certain area but rather several that manifest themselves on different temporal and spatial scales. It could be shown very clearly that alterations within the subsistence strategies in Britain were not merely an adaptation to Roman demands but rather a multifaceted process, which was closely linked to changes in Time-Space Constraints and their implications for trade.

Because the identified outliers are all high status sites, the results do not sustain hypotheses about the emulation of Roman culture by non-élites.

King (1984, 1999) and Hambleton (1999) both used tripolar graphs to analyse faunal remains from Iron Age and Roman Britain. They conclude that the indigenous subsistence strategies in Britain are characterised by sheep as the most abundant species (see section 9.4). The fact that assemblages from British military and Romanized sites show a higher proportion of cattle has been interpreted as an acculturation of the British diet; "In all probability, these trends in diet can be attributed largely to the influence of the army" or "the process of change was probably brought about by emulation of the new elite formed by the Roman army and administration" (King 1984: 190). The results of this case study challenge this understanding. The methods applied allowed a more detailed insight into the British subsistence strategies than tripolar graphs and have revealed that there were already differences between the high status sites prior to the Roman presence. The assemblages of two of the later Romanized group 1 sites (Ditches and Cat's Water) were already dominated by cattle and all the others contained *proportionally* fewer sheep bones. While the presence of the Roman army in Britain might have accelerated an already existing trend, the results of this case study do not support the assumption that they are responsible for it. The high status sites that later became Romanized already showed less focus on sheep husbandry which means that the dietary preferences of the Roman army were less foreign to them and made it easier for them to adapt to their demands.

The discussion on Romanization has argued for the active role of Iron Age communities (Webster 2001). Recent research on the urbanising élite of Iron Age Britain has shown close links between sites in South-East Iron Age Britain and the continent. Pitts & Perring (2006) concluded that the adoption of urban strategies among élite groups in Britain was triggered by their own interests and only partly by Roman incentives. In the same manner this paper has shown that alterations within the subsistence strategies in Britain were not merely an adaptation to Roman demands but rather a multifaceted process, which was closely linked to changes in TS-constraints and their implications for trade. The Romanization of Britain was by no means one-sided or simply forced upon the indigenous population.

I argue that the stronger emphasis on cattle husbandry in the high status sites was not initiated by the Romans (King 1984, 1999) but is the result of a complex interaction with pre-existing Iron Age patterns. This also supports the idea that major changes took place before the Romans settled in Britain (Millett 1990a: 29).

The model introduced in this paper underlines the importance of receptivity in innovative processes. Although starting from a different viewpoint, Cools' pottery based analysis of eating and drinking habits in later Iron Age Britain is entirely consistent with the analysis presented here and equally confirms the important role of cultural aspects; despite trade contacts and geographical proximity to the continent large parts of Iron Age Britain showed no demand for Roman imports, which she concludes "probably lies more in the social than the political realm" (2006: 171). This highlights the role cultural factors take in innovative processes and acceptance of change which is underestimated in much of the economic literature on innovation.

12.3 CONCLUSIONS

The developments in the North-Western Provinces have been described for three different levels; 1) the local level of agricultural strategies represented through the results from the case study, 2) the regional level of trade and economical organization, as well as 3) the imperial level of the Roman Empire represented by warfare and policy strategies of ambitious young men. It has been demonstrated how closely processes on these different levels were connected during the Romanization of the North-Western Provinces. This supports the claim that an explanation of social change should always consider processes manifest on different levels of the social hierarchy as well as a combination of social and physical constraints. This chapter has demonstrated how processes on all three scales have worked to reinforce each other; the demand created on one level needed to be fulfilled through alterations at another level.

It has been argued that the expansion of trade contacts between Gaul and Britain was only possible after long-distance space constraints were overcome. The observed shift in husbandry regimes and economy are inseparable from changes in Time-Space constraints and processes on various scales had to be harmonised in order to allow for the Romanization of Britain. The expansion of the Roman Empire into Gaul triggered the creation of a market economy in some of the British high-status sites. Changes in Gaul reconciled Time-Space constraints acting on Britain and thus allowed for further changes.

This understanding of Romanization as a multiscalar process restricted by the physical landscape can help to explain the failure of Romanization in certain parts of North-West Europe. The proximity of the Roman Empire might have created the same demands in Germany North of the Rhine, Ireland or Scotland but the combination of social and physical constraints did not allow the indigenous population in these areas to react to these demands, even though they are likely to have been increased. These areas might have lacked a redistributive economy which has been proven to be crucial for the developments in Britain; or their location in the landscape has constituted Time-Space constraints that could not have been overcome at the time. This can explain some of the difficulties the Romans had with occupying areas like Ireland, Scotland and parts of Germany. A comparison with these areas could show whether there were significant differences in mode of production, societal hierarchy or the degree of insipient urbanization.

The case study has demonstrated how hypotheses developed in one domain can be explored using another, unrelated set of data. A conceptual model of innovation, based on sociological theory, has been successfully applied to the archaeological record. We have also seen that the same faunal data can be organised in different ways and explored using different statistical methods to tease out information relating to different space-time scales. The method is far from perfect or simple (the data does not speak for itself) but using middle range methods it is sometimes possible to reconstruct a more balanced multi-scalar history than one might expect. Linking faunal data to site-specific information about ceramics, cross-channel trade and off-site information like roads reinforces this multi-scalar picture.

The statistical analysis of the British Iron Age sites has identified two different types of large faunal assemblages. The groups are defined by different sample size, differential preservation and recovery techniques as well as behavioural differences. For the latter the main differential criterion between the two groups seems to be the proportion of sheep in the overall assemblage. While sheep is also the most abundant species in 2 of the 7 group 1 sites, its proportion is significantly higher in the group 2 sites.

If a wider temporal scale is included it becomes apparent that all group 1 sites became Romanized while the group 2 sites did not. Both aspects fit into the literature in as much as the group 2 hillforts represent a 'typical' Iron Age species distribution, with sheep as the most abundant species. Throughout the Roman period the importance of sheep declined in Britain while cattle became more common. At Dragonby and Balksbury the proportion of cattle fragments increases relative to sheep in the Romano-British contexts, even though at Balksbury this could be caused by differential preservation rather than a change in husbandry (Harman 1996, Maltby 1995). The study also confirms that developments compatible with future Romanization were already present in some of the Iron Age assemblages.

13. SUMMARY

This study was concerned with continuity and change, focusing on *innovation* as one mechanism of social change. The literature revealed that - despite using the same term - different disciplines of the social sciences have very different understandings of what innovation actually is. Economists consider it to be an important factor in economic growth and focus almost exclusively on technological development. Politicians express a very similar view; both are orientated towards the future and have a strong wish to manage innovative processes. Archaeologists, on the other hand, focus on cultural aspects, like the role of social acceptance and individual perception. Cultural Geographers study the diffusion of innovation as a spatial process. Thus an initial literature review has revealed that, despite the consensus that innovation is important, no uniform definition exists.

This is not surprising considering that innovation is a nonlinear, complex social process. The problems of the social sciences are often 'messy' in as much as no time-invariant objective solution exists for them. This is because human knowledge is socially constructed and, as that knowledge changes, human behaviour and perception actually change with it. Hence the first part of this book developed a *conceptual model* of innovation as a basis for interdisciplinary discussion and research. Modelling provides a valuable tool to simplify and operationalise complex social phenomena, because it forces the researcher to identify the crucial components in regard to the research question.

One common feature within the diversity of the innovation literature is the importance attributed to *knowledge*. Chapter 4 has argued that knowledge in respect to innovation should be seen as shared beliefs rather than pure technological expertise. While the latter perception dominates the economic and political understanding, it is *qualitatively* different, rather than just quantitatively more, information that is crucial to innovative processes. Based on a sociological understanding of the human condition a conceptual model was developed that defines innovation as "an information flow that challenges beliefs and leads to a change of behaviour" (Winder 2005: 275). The term 'beliefs' describes the internalised categories and presumptions that build the basis of every human action and determine the way we perceive and evaluate the outside world. Innovation is seen as a process of *reconceptualisation.*

This definition binds the starting point of every innovation to the micro level, because it is always an individual that receives and processes challenging information from its environment. While the cause of innovation is bound to an individual its observable effects - a change in behaviour - can diffuse and become a macro phenomenon. Many disciplines that are concerned about innovation focus exclusively on macro-level changes.

Every model should be tested in the data. This presents a challenge because changes in beliefs are not observable in the archaeological record. This book has introduced the concept of *Time-Space constraints* as a proxy measure. They represent the impact of the environment on each individual. Changes in TS constraints suggest that the probability of reconceptualisation rises. The human condition is determined by two processes; collective human action can alter the structures acting upon them from the *bottom up* and at the same time these structures pose opportunities or threats that influence the adaptive potential of each individual from the *top down.* Changes in top-down processes are experienced by the affected individual as changes in the environment. The majority of human action is based on routines and habitus, which are likely to become ineffective within a changing environment. Thus, by creating new conditions for people to act in, changes in Time-Space constraints are likely to make individuals see opportunities or threats they were not able to visualise before. The relationship between Time-Space constraints and a challenge of beliefs is by no means deterministic but rather one of probabilities; changes in TS constraints *increase the likelihood* of innovation occurring.

The model was tested in the archaeological record with an analysis of the Romanisation of the North-Western Provinces from 50 BC to AD 50. This choice of case study was a rather random one, because the model should be applicable to all innovations. Thus the resulting case study differs from other approaches to Romanization in as much as it departs from a specialised disciplinary perspective.

The faunal remains of sheep, cow and pig of 46 and of 37 British Iron Age sites, respectively, were analysed with visual techniques as well as a Principal Component Analysis. Both methods identified two outlier groups which contained all the large assemblages from the database. The differences between both groups are partly caused by taphonomic processes and differential recovery techniques but they also reflect differences in subsistence strategies and thus human behaviour. This allows a connection to the conceptual model. The behavioural differences between the outliers manifest themselves in a proportionally higher sheep distribution in one group. The shift to another temporal scale revealed that these high-status sites experienced a disruption in occupation and declined while sites in the other group became Romanized. Thus the analysis revealed two different types of Pre-Roman high-status sites: one that later became Romanized and one that did not. It is remarkable that this result was gained from Iron Age remains and thus with an *a priori perspective* rather than as a *retrospective* observation.

This pre-Roman variability, which has been interpreted in terms of mixed arable farming with more cattle as opposed to animal husbandry with more sheep, split large assemblages with demographic and sociological conditions in favour of Romanization into those that persisted into the Roman period and those that dwindled.

13.1 PROSPECTS

As a result of this research some suggestions can be made regarding future research, both for archaeology in particular as well as for the development and the use of conceptual models in the social sciences in general.
The data analysis of faunal remains has revealed several issues in regard to the quantification and documentation of archaeological data. The study has shown once more the need to identify taphonomic processes before any observed pattern is interpreted in regard to human behaviour and chapter 11 has given an example of how the effects of these potential biases can be identified relatively easily. Any study of faunal remains should include similar tests to minimize the risk of misinterpretation.

This research has encountered several issues with the quantification of faunal remains. The site reports, consulted to analyse the outlier differences, have demonstrated scale and quantification method dependencies of faunal remains. The report for Ditches hillfort, for instance, contains two reports on animal bones, one by Rielly & Trow (1988) and one by Rielly (1988). Both give a graphical representation of the relative species proportions and show crucial differences: while cattle is the most abundant species in the histograms (46%, followed by sheep with 38%) sheep dominates in the tripolar graph (43%, followed by cattle with 37%). The tripolar representation contains only these three domesticates while the histograms also account for dog, horse and deer bones found at the site. However, even the corrected histograms[1] do not level out these discrepancies. There is no suitable explanation for these differences within the reports, but it is likely that the exclusion of ambiguous material from one of the databases, the collapsing of data categories or the representation of different temporal or contextual layers could account for some of the dissimilarities. In the overall site assemblage (Rielly 1988), Ditches is one of the few outlier sites with cattle as the most abundant species; however this pattern is not confirmed by the tripolar representation of the data. The species distribution has also been shown to be dependent on the quantification method; cattle bones are less well represented in the MNI counts due to a proportionally high fragmentation in this species (Rielly 1988).

As another example, the bone frequencies for Balksbury (MIA) listed in Hambleton (1990: appendix 3) show great discrepancies from the animal bone site report based on the AML Lab report no 2918 (Maltby 1995) (see table 23). This can be explained with the fact that Hambleton extracted her data directly from the AML report rather than its summary (pers. comm.). The differences between the two documents are likely to be caused by the exclusion of Roman or ambiguous material (Maltby, pers. comm.).

[1] By excluding the other animals and spreading the three main domesticates over 100%.

Both cases show how different sources for the same site might differ significantly in their representation of faunal remains, providing a different picture of its subsistence strategies. This provides a good example for the impact of publication decisions on our understanding of archaeological artefacts.

Element	*Maltby 1995* middle to late Iron Age	*Hambleton 1990* middle Iron Age
Cattle		
Mandible	205	217
Phalanges	49	19
Vertebrae	81	68
ribs	6	6
Sheep		
Mandible	273	274
Phalanges	56	18
Vertebrae	280	157
ribs	138	0

***Table 23**: Differences in species frequencies for Balksbury Camp*

These examples demonstrate that standards for the quantification of artefacts are vital and some of the earlier reports might need repeating once such standards are in place.

Despite all these problems it has been argued that archaeologists should avoid considering the archaeological record as too biased to draw any sensible conclusions about the past; these limitations should not lead to a rejection of empirical studies. This claim is supported by the successful quantitative analysis of faunal remains introduced in this book, despite an array of 'biases' coded into the assemblages.

The conceptual model has been successfully applied to the archaeological record and hypotheses developed independently under a sociological framework could be tested. This approach is unusual for archaeologists but has proven to be very fruitful. This should encourage archaeologists to engage in this method more often, since it provides protection from circular argumentation which constitutes a common problem within the discipline. The initial hypothesis independently confirmed commonly accepted theories of Romanization, which gives them additional power.

It has been shown how processes acting on different temporal and spatial scales had to interact to create the transition from the Late Iron Age to the Roman period in Britain. This study only considered patterns in subsistence strategies represented by the three main domesticates, sheep, cow and pig, and an analysis based on other artefacts could benefit our understanding of these processes.

Some of the big oppida in the South East of Britain could not be included in this study due to their lack of faunal data. Large high-status settlements played a key role during the Romanization of Britain. A more complete dataset based on other artefacts and thus capable of including these missing sites would provide us with a more detailed understanding of the processes in this area.

This study did not include faunal remains of areas that resisted Romanization. The last chapter has provided hypotheses as to why these areas did not become Romanized and suggested that processes on various scales could not be harmonized due to rigid Time-Space constraints. It would be interesting to see which patterns and trends in subsistence strategies can be identified for Ireland or Germany east of the Rhine. The existence of hillforts in Ireland shows that the social organisation and economy of these people was efficient enough to construct such large fortified settlements. However these societies might have differed from Britain in two ways 1) the economy had not necessarily developed as far as a redistributive system and 2) the use of these settlements might have been very different. It has been suggested that the hillforts of Ireland mainly provided temporary refuge in times of danger and did not show signs of development towards early urbanization (Raftery 1972: 39). Such an inter-regional comparison is likely to provide more detailed insight into the factors that constitute whether an innovation is accepted or rejected.

Due to the high numbers of missing values on the skeletal element information only 1 of the 6 Scottish sites included in Hambleton's study (Port Seaton) could be considered in this research. A better understanding of the similarities or differences in Iron Age husbandry regimes north of Hadrian's Wall would allow us to test whether the pattern identified for the South is applicable to the entire country. Due to its close connection with the hierarchal organization at a site, the mode of production promises to be an interesting factor to test for. Climate and geography provide very different conditions in Scotland and it is assumed that fish and shellfish were used to compensate for the disadvantages the environment provided for agriculture and animal husbandry. Cunliffe (2000: chapter 15) has argued that settlements in this region were able to provide for themselves but did not produce a surplus for trade like the South. Thus the Pre-Roman husbandry regimes in Scotland are likely to have differed from the practices in the group 1 sites and possibly also from Danebury and Maiden Castle. An analysis of Scottish faunal remains would therefore not only allow the comparison of sites that became Romanized and sites that did not, but also between two different types of settlements that rejected Romanization.

13.2 RELEVANCE OF THIS RESEARCH TO A WIDER COMMUNITY

Archaeology is *"the discipline with the theory and practice for the recovery of unobservable hominid behaviour patterns from indirect traces in bad samples"* (Clarke 1979: 100)

Of all the social sciences archaeology is unusual in having no direct access to human behaviour and its underlying intentions. It has been argued that different aspects of human societies vary in the degree to which they are represented in the archaeological record; technology or economic behaviour is easier to reconstruct than beliefs or social organisation. The reconstruction of these latter, key aspects of human interaction from static artefacts remains one of the main issues in archaeology, ever since Clarke's 1973 paper 'Archaeology: the loss of innocence' made explicit the need for ways of connecting archaeological remains to human behaviour explicit. The last century has seen various approaches to solving these difficulties (see chapter 5). Empirical archaeology describes patterns identified in quantified data and generates typologies. New or Processual archaeology uses experimental archaeology and anthropological studies to gain a better understanding of the circumstances under which the archaeological record developed. This paradigm claims that the correct use of scientific methods could overcome the downfalls of the archaeological record (Binford 1983). Post-Processual archaeology concentrates on discursive methods. It argues that artefacts were 'meaningfully constituted' and that the role of beliefs is therefore crucial for our understanding of the archaeological record. Because meaning is observer-dependent archaeology can never develop time-invariant 'laws' about the past (Hodder 2005).

The methods applied to this case study do not exclusively fit into any of these paradigms but rather constitute a mixture of all of them; this research has combined empirical data with a discursive model based on the sociological understanding of the human condition in order to gain insight into the behaviour of people living in the British Iron Age. Its statistical analyses fall into the tradition of empiricism. At the same time the exploratory approach using various visual methods has shown how approaching the same data set from a variety of angles can result in different pictures about the past. The distribution of skeletal elements, for instance, provides information on a level below an inter-site comparison of species distribution. And connecting the empirical study to Romanization was only possible after a deeper temporal scale was consulted. The problem of inferring information about human beliefs from archaeological artefacts was approached with middle range theory as suggested by Binford. The combination of sociological theory with concepts developed in Time-Geography and TiGrESS introduced the concept of Time-Space constraints as a means of understanding the archaeological record. This allows the inclusion of human-modified landscapes into the understanding of social change.

TiGrESS (Winder 2006) has provided us with various successful case studies that managed to combine quantitative and discursive methods in policy-related areas. This book has combined both methods for archaeology and used them to monitor processes on different temporal and spatial scales. Rather than seeing them as low-dimensional artefacts, the faunal remains were considered to provide complex information about the past. This study approached the transition from the British Iron Age to the Roman period by combining data from various different scales and analysing their interaction. Future research should be encouraged to look simultaneously at different spatial and temporal scales because it has been demonstrated quite clearly that innovation happens simultaneously in all of them.

One aspect that reappeared throughout the course of this research is the *impact of observation scale and length* on the observed pattern. Its influence was discussed in three different but related contexts:

1) The Simpson's paradox (see section 8.2) implies that the collapsing of categories might produce different patterns from grouped data. This suggests that archaeological data can never be interpreted objectively, because we have no way of telling which pattern is right and which pattern is wrong. Simpson's paradox is valid for all forms of data. The social sciences have no fixed 'units of data' that determine the scale of observation. It is rather the research question that defines the unit and scale of observation. Sociologists for instance can study the problem of unemployment on a local level, using individuals as the unit of observation or on a global level concentrating on the relationship between multi-national companies.

2) The interpretation of the pattern found in Iron Age faunal remains has demonstrated how valuable the perspective of the *longue durée* (Braudel 1980) can be for the understanding of social change. Observation on different time scales provide very different information and processes that are not visible during the *l'histoire événementielle* (the short term observation of single events) will become visible in the *longue durée*. The damaging effects of smoking for instance only become visible after a longer time period and any short-term study would not be able to detect them. Despite all the disadvantages of the archaeological record discussed above its access to deep time gives it a significant advantage for the study of social change. Archaeological studies may help to complement results gained in other social disciplines that focus on the future rather than the past and deal with considerable shorter time spans.

3) The first part of this book has shown that innovation is a nonlinear, complex social phenomenon whose effects might become manifest on several temporal and spatial scales. The conceptual model explicitly aims to accommodate this diversity by considering both the cause of innovation (manifest on the micro level) and its effects which might become visible on an aggregated level of society. It thus allows the observation of innovation at different scales.

Finally, the results of this case study should encourage us to think differently about innovation in disciplines other than archaeology. Economical and political approaches to innovation with a strong emphasis on technological expertise should consider widening their understanding of knowledge. This book has argued strongly that innovation cannot be equated with technological development or economic growth but is a process of reconceptualisation. The definition of knowledge as information that allows us to see new opportunities or threats might help to explain some of the problems experienced with transferring R&D into exploitable products.

Most of the literature on innovation makes us believe that innovation is a 'good thing'. Economists and Politicians equate it with economic growth and an essential advantage in market competition. "It is almost universally accepted that technological change and other kinds of innovations are the *most important sources of productivity growth* and *increased material welfare* – and that this has been so for centuries" (Edquist 2000: 3). West and Farr define innovation as "the intentional introduction and application within a role, group or organization of ideas, processes, products or procedures, new to the relevant unit of adoption, designed to *significantly benefit the individual, the group, organization or wider society*" (1990: 9). This book has shown that humans tend to be reluctant regarding change. The majority of human behaviour is based on routine and habitus, the exact opposite of innovation. This makes perfect sense in regard to human interaction with others and the environment; any society relies on a certain amount of stability. Innovation as the recognition of new opportunities or threats can be painful or create stress because strategies that have worked in the past are no longer valid. Some innovations also alter the processes within a system. Innovations can push our knowledge about the world past its sell-by date. This is not to say that innovation is not crucial or beneficial, but it should not be forgotten that it often comes at a price, the price of unpredictability and uncertainty.

BIBLIOGRAPHY

Acs ZJ, Audresch DB & Feldman MP (1992) "Real Effects of Academic Research: A Comment." *The American Economic Review* **82**(1): 363-367.

Ahl V & Allen TFH (1996). *Hierarchy Theory. A Vision, Vocabulary, and Epistemology*. New York, Columbia University Press.

Ahn S (2002) Competition, Innovation and Productivity Growth: A Review of Theory and Evidence and Evidence. Economics Department Working Papers No. 317. Paris. OECD.

Allen PM (1989) Modelling innovation and change. In: van der Leeuw SE & Torrence R *What's New? A Closer Look at the Process of Innovation*. Boston, Unwin Hyman Ltd: 258-280.

Allen PM & McGlade JM (1986) "Dynamics of discovery and Exploitation: The Case of the Scotian Shelf Groundfish Fisheries." *Canadian Journal of Fisheries and Aquatic Sciences* **43**: 1187-1200.

Allen PM & McGlade JM (1987) "Modelling complex human systems: A fisheries example." *European Journal of Operational Research* **30**: 147-167.

Ammerman AJ & Cavalli-Sforza LL (1971) "Measuring the Rate of Spread of Early Farming in Europe." *Man* **6**(4): 674-688.

Ammerman AJ & Cavalli-Sforza LL (1973) A population model for the diffusion of early farming in Europe. *The explanation of culture change: models in prehistory*. Renfrew C, Duckworth: 343 - 357.

Archer MS (1982) "Morphogenesis versus Structuration: On Combining Structure and Action." *The British Journal of Sociology* **33**(4): 455-483.

Archer MS (2000) *Being Human: the Problem of Agency*. Cambridge, Cambridge University Press.

Archer MS (2005) Structure, Culture and Agency. In: Jacobs MD & Weiss Hanrahan N *The Blackwell Companion to the Sociology of Culture*. Oxford, Blackwell: 17- 34.

Armour-Chelu M (1991) The faunal remains. In: Sharples NM *Maiden Castle: Excavations and Field Survey 1985-1986*. London, English Heritage Archaeological Report 19: 139-151.

Asch SE (1958) Effects of group pressure upon the modification and distortion of judgments. In: Maccoby EE, Newcomb TM & Hartley EL *Readings in Social Psychology*. New York, Holt, Rinehart & Winston: 174-183.

Asheim BT & Gertker MS (2005). The Geography of Innovation. Regional Innovation Systems. In: Fagerberg J, Mowery DC & Nelson RR *The Oxford Handbook of Innovation*. Oxford, Oxford University Press: 291-317.

Audretsch DB & Feldman MP (1996) "R&D Spillovers and the Geography of Innovation and Production." *The American Economic Review* **86**(3): 630-640.

Bargatzky T (1989) Innovation and the integration of sociocultural systems. In: van der Leeuw S & Torrence R *What's New? A closer Look at the Process of Innovation*. London, Unwin Hyman: 16-32.

Barrett J & Fewster K (2000) Intimacy and Structural Transformation. In: Holtorf C & Karlsson H *Philosophy and Archaeological Practice. Perspectives for the 21th Century*. Göteberg, Bricoleur Press: 25-38.

Basberg BL (1987) "Patents and the measurement of technological change: A survey of the literature." *Research Policy* **16**: 131-141.

BCG (2003) Raising the Return on Innovation. Innovation-to-cash survey 2003, The Boston Consulting Group. *http://www.bcg.com/publications/files/ITC_Survey_rpt__Dec03.pdf*, accessed January 2004

Becker G (1981) *Treaties in the Family*. Cambridge, Harvard University Press.

Benninghaus H (1998) *Deskriptive Statistik*. Stuttgart, Teubner.

Berry B & J Kasarda (1977) *Contemporary Urban Ecology*. New York, Macmillian Publ. Co.

Biddick K (1984) Animal Bones from Cats Water subsite. In: Pryor F *Excavations at Fengate, Peterborough, England: The fourth report*. Northamptonshire Archaeological Society Monograph 2: Michrofiche: 245-275.

Binford LR (1965) "Archaeological Systematics and the Study of Cultural Processes." *American Antiquity* **31**: 203-210.

Binford LR (1978) *Nunamiut Ethnoarchaeology*. New York, Academic Press.

Binford LR (1981) *Bones. Ancient men and modern myths*. New York, Academic Press.

Binford LR (1983) *In Pursuit of the Past. Decoding the archaeological record*. German Democratic Republic, Thames & Hudson.

Bonabeau E, Dorigo M & Theraulaz G (1999) *Swarm intelligence: from natural to artifical systems*. New York, Oxford University Press.

Bourdieu P (1983) Ökonomisches Kapital, kulturelles Kapital, soziales Kaptital. *Soziale Ungleichheiten*. Kreckel R. Göttingen, Schwartz: 183-198.

Bourdieu P (1986) The Forms of Capital. In: Richardson JE *Handbook of Theory and Research for the Sociology of Education*. Westport, Greenwood Press: 241-258.

Boyd R & Richerson PJ (1985) *Culture and the evolutionary process*. Chicago, University of Chicago Press.

Brain CK (1981) *The Hunters or the Hunted? An Introduction to African Cave Taphonomy*. Chicago, University of Chicago Press.

Braudel F (1980) *On History*. London, Weidenfeld and Nicolson.

Bretagnolle A, Paulus F & Pumain D (2005) *Networks and urban scaling*. Paper given at conference "Urban scaling and innovation" Santa Fe Institute (New Mexico) July 23 – August 1.

Caesar (1966) *The Gallic War. With an English Translation by HJ Edwards*. London, William Heineman Ltd.

Carlstein T (1978) Innovation, Time Allocation and Time-Space Packing. In: Carlstein T, Parkes D & Thrift N *Human Activity and Time Geography, Volume 2*. Bungay, Arnold. **2:** 146-161.

Carroll M (2001) *Romans, Celts & Germans. The German Provinces of Rome*. Stroud, Tempus Publishing Ltd.

Chaplin RE (1971) *The Study of Animal Bones from Archaeological Sites*. London, Seminar Press.

Checkland P (1991) From Optimizing to Learning: A Development of Systems Thinking for the 1990s. In: Flood RL & Jackson MC *Critical Systems Thinking*. Cambridge, Wiley: 59-75.

Checkland P (1993) *System Thinking, Systems Practice*. Chichester, John Wiley & Sons.

Checkland P (2001) Soft System Methodology. In: Rosenhead J & Mingers J *Rational Analysis for a Problematic World Revisited. Problem Structuring Methods for Complexity, Uncertainty and Conflict*. Chichester, John Wiley & Sons: 61-89.

Checkland PB & Casar A (1986) "Vickers' concept of an appreciative system: A systematic account." *Journal of Applied Systems Analysis* **13**: 2-17.

Checkland P & Howell S (1998) *Information, Systems and Information Systems*. Chichester, John Wiley & Sons.

Chevallier R (1976) *Roman Roads*. London, Batsford Ltd.

Child AM (1995) "Towards an understanding of the microbial decomposition of archaeological bone in the burial environment." *Journal of Archaeological Science* **22:** 165-174.

Childe VG (1929) *The Danube in prehistory*. Oxford, The Clarendon Press.

Clarke DL (1972) Models and paradigms in contemporary archaeology. In: Clarke DL *Models in Archaeology*. London, William Cloves & Sons: 1-60.

Clarke DL (1973) "Archaeology: the loss of innocence." *Antiquity* **XLVII**: 6-18.

Clarke DL (1979) *Analytical Archaeologist*. New York, Academic Press.

Cohen W & Levin RC (1989) Empirical Studies of Innovation and Market Structure. In: Schmalensee R & Willig RD *Handbook of Industrial Organization, Volume 2*. Amsterdam, North Holland. **2:** 1059-1107.

Collingwood RG (1946) *The Idea of History*. Oxford, Oxford University Press.

Collis J (1982) Gradual growth and sudden change - urbanisation in temperate Europe. In: Renfrew C & Shennan S *Ranking, resource and exchange*. Cambridge, Cambridge University Press: 73-78.

Cool HEM (2006) *Eating and Drinking in Roman Britain*. Cambridge: Cambridge University Press.

Coy J (1987) The animal bones and marine molluscs. In: Woodward PJ *Romano-British Industries in Purbeck. Excavations at Ower and Rope Lake Hole*. Dorset Natural History and Archaeological Society Monograph 6: 177-179.

Cunliffe B (1976) The origins of urbanisation in Britain. In: Cunliffe B & Rowley T *Oppida: the beginning of urbanization in Babarian Europe*. Oxford, B.A.R. publications: 135-163.

Cunliffe B (1978) *Rome and Her Empire*. Ludwigsburg, Süddeutsche Verlagsanstalt.

Cunliffe B (1984) Community, continuity, and change. In: Cunliffe BW *Danebury. An Iron Age Hillfort in Hampshire. Vol 2 The excavations 1969-1978: the finds*. CBA Research Report No 52: 549-559.

Cunliffe B (2000) *Iron Age Communities in Britain.* London, Routledge.

Darwin C (1859) *On the Origins of Species*. London, John Murray.

David PA (1985) "Clio and the Economics of QWERTY." *The American Economic Review* **75**(2): 332-337.

Davis SJM (1995) *The Archaeology of Animals*. London, Batsford Ltd.

Dawe A (1970) "The Two Sociologies." *The British Journal of Sociology* **21**(2): 207-218.

Dawkins R (1976) *The Selfish Gene*. Oxford, Oxford University Press.

Dean SD, Gumerman GJ, Epstein JM, Axtell RL, Swedlund AC, Parker MT & McCarroll S (2000) Understanding Anasazi Culture Change Through Agent Based Modeling. In: Kohler TA & Gumerman GJ *Dynamics in Human and Primate Societies*: 179-205.

Dennett D (1991) *Consciousness explained.* London, Penguine Press.

Dosi G (2000) The Research on Innovation Diffusion: An Assesment. In: Dosi G *Innovation, Organization and Economic Dynamics. Selected Essays.* Cheltenham, Edward Elgar Publishing: 115-144.

Drinkwater JF (1983) *Roman Gaul*. London, Canberra, Croom Helm.

Durkheim E (1885) *Regeln der soziologischen Methode.* Darmstadt/ Neuwied, Luchterhand.

Durkheim E (1950) *The Rules of Sociological Method.* New York, The Free Press.

Dyer J (1992) *Hillforts of England and Wales*, Shire Publications LTD.

Edmondson JC (1990) Romanization and Urban Development in Lusitania. In: Blagg T & Millett M *The Early Roman Empire in the West.* Oxford, Oxbow Books: 151-178.

Edquist C (2000) Systems of Innovation Approaches - Their Emergence and Characteristics. In: Edquist C & McKelvey M *Systems of Innovation: Growth, Competitiveness and Employment. Volume 1.* Cheltenham, Elgar Reference Collection: 3-37.

Ellen R (1994) Rates of Change: Weasel Words and the Indispensable in Anthropological Analysis. In: Hann CM *When History Accelerates. Essays on Rapid Social Change, Complexity and Creativity*. London, The Athlone Press: 54-73.

Esser H (1990) ""Habitus", "Frames" und Rational Choice. Die Reichweite der Theorie der rationalen Wahl (am Beispiel der Erklärung des Befragtenverhaltens)." *Zeitschrift für Soziologie* **19**: 231-247.

Esser H (1993) *Soziologie – Allgemeine Grundlagen.* Frankfurt, Campus.

EU (1995) Green Paper on Innovation. Accessed January 2004 *http://ec.europa.eu/comm/off/pdf/COMM(95)688 .pdf*

EU Policy Committee (2005) Report on the Lisbon National Reform Programmes 2005. Brussels, EU. Accessed January 2004 *http://ec.europa.eu/economy_finance/epc/epc_p ubliscations_en.htm#lisbon*

Fagerberg J (2005) Innovation. A guide to the literature. In. Fagerberg J, Mowery DC & Nelson RR *The Oxford Handbook of Innovation.* Oxford, Oxford University Press: 1-26.

Favory F, Girardot JJ & Raynaud C (2003) Typologie archaeologique de l'habitat rural. In : van der Leeuw S, Favory F & Fiches JL *Archaeologie et systemes socio-environmentaux. Etudes multiscalaires sur la vallee du Rhone dans le programme Archaeomedes.* Paris, CNRS Editions: 201-238.

Feldman M (1994) *The Geography of Innovation.* Dordrecht, Kluwer Academic.

Fifield PW (1988) "The faunal remains." *Hertfordshire Archaeology,* **10**: 148-153.

Fitzhugh B (2001) "Risk and Invention in Human Technological Evolution." *Journal of Anthropological Archaeology* **20**: 125-167.

Fitzpatrick A & Timby J (2002) Roman Pottery in Iron Age Britain. In: Woodward A & Hill JD *Prehistoric Britain. The Ceramic Basis.* Oxford, Oxbow Books: 161-172.

Flam H (1990) "Emotional 'Man': I. The Emotional 'Man' and the Problem of Collective Action." *International Sociology* **5**(1): 39-56.

Flannery K (1968) Archaeological system theory and early Mesoamerica. In: Meggers BJ

Anthropological Archaeology in the Americas: 67-87.

Fliegel FC & Kivlin JE (1966) "Attributes of Innovations as Factors in Diffusion." *The American Journal of Sociology* **72**(3): 235-248.

Freeman C (1982) *The Economics of Industrial Innovation.* London, Frances Pinter.

Freeman C & Soefe L (1997) *The Economics of Industrial Innovation.* London, Continuum.

Frere S (1987) *Britannia: a history of Roman Britain.* London, Routledge & Kegal Paul.

Garfinkel H (1973) Das Alltagswissen über und innerhalb sozialer Strukturen. *Alltagswissen, Interaktion und gesellschaftliche Wirklichkeit.* Arbeitsgruppe Bielefelder Soziologen, Rowohlt. **1:** 189-262.

Giddens A (1984) *The Constitution of Society.* Cambridge, Polity Press.

Giddens A & C Pierson (1998) *Conversations with Anthony Giddens.* Cambridge, Polity Press.

Goffman E (1974) *Frame Analysis. An Essay on the Organization of Experience.* Cambridge, Massachusetts, Harvard University Press.

Grant A (1984a) Animal Bones. In: Cunliffe BW *Danebury: an Iron Age hillfort in Hampshire. Volume 2: the excavations 1969-78: the finds.* London, CBA. **2:** 496-547.

Grant A (1984b) Survival or sacrifice? A critical appraisal of animal burials in Britain in the Iron Age. In: Grigson C & Clutton-Brock J *Animals and Archaeology 4: Husbandry in Europe.* Oxford, BAR International Series. **227:** 221-227.

Grant A (1991) Animal Bones. In: Cunliffe BW & Poole C *Danebury: an Iron Age hillfort in Hampshire. Volume 5: the excavations 1979-88: the finds.* London, CBA. **5:** 47-87.

Grayson DK (1973) "On the methodology of faunal analysis." *American Antiquity* **39**(4): 432-439.

Grayson DK (1978) "Minimum Numbers and Sample Size in Vertebrate Faunal Analysis." *American Antiquity* **43**(1): 53-65.

Greene K (1978) Roman trade between Britain and the Rhine provinces: the evidence of pottery to c AD 250. In: du Plat Taylor J & Cleere H *Roman shipping and trade: Britain and the Rhine provinces*. London, CBA report. **24:** 52-58.

Greene K (1986) *The Archaeology of the Roman Economy.* Berkeley, University of California Press.

Griliches Z (1990) "Patent Statistics as Economic Indicators: A Survey." *Journal of Economic Surveys* **28**(4): 1661-1707.

Guille-Escuret G (1993) Technical Innovation and Cultural Resistance. In: Lemonnier P *Technological Choices. Tranformation in material cultures since the Neolithic.* London, Routledge**:** 214-226.

Hägerstrand T (1967) *Innovation Diffusion as a spatial process.* Chicago, University of Chicago Press.

Hägerstrand T (1975) Space, Time and Human Conditions. In: Karlqvist A, Lundqvist L & Snickars F *Dynamic Allocation of Urban Space.* Lexington, Lexington Books: 3-14.

Hägerstrand T (1987) On Levels of Understanding the Impact of Innovations. In: Fischer M, Sauberer M *Gesellschaft, Wirtschaft, Raum. Beiträge zur modernen Wirtschafts- und Sozialgeographie. Festschrift für Karl Stiglbauer.* Wien, Arbeitskreis für Neue Methoden in der Regionalforschung**:** 46-50.

Hägerstrand T (1988) Some Unexplored Problems in the Modelling of Culture Transfer and Transformation. In: Hugill PJ & Dickson DB *The Transfer and Transformation of Ideas and Material Culture.* Texas, Texas A&M University Press**:** 217-232.

Haken H (1978) *Synergetics: An Introduction.* Berlin, Springer Verlag.

Hall BH (2005) Innovation and Diffusion. In: Fagerberg J, DC Mowery DC & Nelson RR *The Oxford Handbook of Innovation.* Oxford, Oxford University Press**:** 459-484.

Hambleton E (1999) *Animal Husbandry Regimes in Iron Age Britain. A comparative study of faunal assemblages from British Iron Age sites.* BAR Series 282, Oxford, BAR Publishing.

Hamilton EG (1996) *Technology and Social Change in Belgic Gaul: Copper Working at the Titelberg, Luxembourg, 125 BC-300AD.* Philadelphia, University of Pennsylvania Museum of Archaeology and Anthropology.

Hamilton WD (1964) "The genetic evolution of social behaviour." *Journal of Theoretical Biology* (7): 1-52.

Hann CM (1994) *When History Accelerates. Essays on Rapid Social Change, Complexity and Creativity*. London, The Athlone Press.

Harman M (1996) Animal remains. In: May J *Dragonby: Report on excavations at an Iron Age and Romano-British settlement in North Linconshire*. Oxford, Short Run Press. **1:** 141-171.

Hartwig F & Dearing BE (1979) *Exploratory Data Analysis*. Beverly Hills, Sage University Papers.

Hassall M (1978) Britain and the Rhine provinces: epigraphic evidence for Roman trade. In: du Plat Taylor J & Cleere H *Roman shipping and trade: Britain and the Rhine provinces*. London, CBA report. **24:** 41-48.

Hill JD (1995) "The Iron Age in Britain and Ireland (c. 800 BC to AD 100): An overview." *Journal of World Prehistory* **9**: 47-98.

Hill JD (2002a) Pottery and the Expression of Society, Economy and Culture. In: Woodward A & Hill JD *Prehistoric Britain. The Ceramic Basis*. Oxford, Oxbow Books: 75-84.

Hill JD (2002b) Just about the Potter's Wheel? Using, making and Depositing Middle and Late Iron Age Pots in East Anglia. In: Woodward A & Hill JD *Prehistoric Britain. The Ceramic Basis*. Oxford, Oxbow Books: 143-160.

Hodder I (1995) *Theory and Practice in Archaeology*. London, Routledge.

Hodder I (2005) Post-processual and interpretive archaeology. In: Renfrew C & Bahn P *Archaeology. The Key Concepts*. Oxford, Routledge: 207-211.

Holland JH (1995) *Hidden Order. How Adaptation Builds Complexity*. Massachusetts, Helix Books.

House of Lords Select Committee on Science and Technology (1997). *The Innovation-Exploitation Barrier*. Third Report, HL Paper 62, March 1997.

Howells J & Roberts J (2000) "From Innovation Systems to Knowledge Systems." *Prometheus* **18**(1): 17-31.

Hradil S (1992) System und Akteur. Eine empirische Kritik der soziologischen Kulturtheorie Bourdieus. In: Müller HP *Sozialstruktur und Lebensstile: der neue theoretische Diskurs über soziale Ungleichheit*. Frankfurt, Suhrkamp: 111-141.

Jaffe AB (1986) "Technological Opportunity and Spillovers of R&D: Evidence from Firms' Patents, Profits, and Market Value." *The American Economic Review* **76**(5): 984-1001.

Janis I & Mann L (1977) *Decision Making. A Psychological Analysis of Conflict, Choice, and Commitment*. New York, The Free Press.

Johnson M (1999) *Archaeological Theory. An Introduction*. Oxford, Blackwell Publishers.

Johnston RJ, Gregory D, Pratt G & Watts M (2000). *The Dictionary of Human Geography*. Malden, Blackwell Publishers.

Jolliffe IT (2002) *Principal Component Analysis*. New York, Springer.

Jones B & Mattingly D (1990). *An Atlas of Roman Britain*. Oxford, Oxbow Books.

King AC (1978) "A comparative survey of bone assemblages from Roman sites in Britain." *Bulletin Institute of Archaeology University of London*(15): 207-232.

King AC (1988) "A comment on the animal bones." *Hertfordshire Archaeology* **10**: 154-155.

King AC (1991) Food production and consumption - meat. In: Jones RFJ *Britain in the Roman Period: recent trends*. Sheffield, R Collis: 15-20.

King AC (1999) "Diet in the Roman World: A regional inter-site comparison of the mammal bones." *Journal of Roman Archaeology* **12**: 168-202.

Klein RG (1980) The Interpretation of Mammalian Faunas from Stone Age Archaeological Sites, with Special Reference to Sites in the Southern Cape Province, South Africa. In: Behrensmeyer AK & Hill AP *Fossils in the Making*. Chicago, University of Chicago Press: 223-246.

Kline SJ & Rosenberg N (1986) An Overview of Innovation. In: Landau R & Rosenberg N *The Positive Sum Strategy: Harnessing Technology for Economic Growth*. Washington DC, National Academy Press: 275-304.

Koestler A (1967) *The ghost in the machine*. London, Hutchinson.

Kohler TA (2000) Putting Social Sciences Together Again: An Introduction to the Volume. In: Kohler TA & Gumersman GS *Dynamics in Human and Primate Societies*. New York, Oxford University Press: 1–18.

Kuhn T (1996) *The Structure of Scientific Revolutions*. Chicago, University of Chicago.

Lemonnier P (1993) Introduction. In: Lemonnier P *Technological Choices. Transformation in material cultures since the Neolithic*. London, Routledge: 1-35.

Maltby M (1987) The animal bones from the excavations at Owslebury, Hants. An Iron Age and Early Romano-British settlement. London, Ancient Monuments Laboratory Report 6/87.

Maltby M (1994) Animal Exploitation in Iron Age Wessex. In: Fitzpatrick AP & Morris EL *The Iron Age in Wessex: Recent Work*. Dorchester, Dorset Press: 9-10.

Maltby M (1995) Animal bone. In: Wainwright GJ & Davies SM *Balksbury Camp Hampshire. Excavations 1973 and 1981*. London, English Heritage. **Report 4:** 83-87.

March J & Simon HA (1958) *Organizations*. New York, Wiley.

Marx K (1964) *Selected Writings in Sociology and Social Philosophy*. London, McGraw Hill.

May J (1996) The Iron Age and Romano-British Settlement. In: May J *Dragonby. Report on Excavations at an Iron Age and Romano-British Settlement in North Lincolnshire*. Oxford, Oxbow. **1:** 69-130.

McCarthy MR (1991) *Roman waterlogged remains at Castle Street. Carlisle: excavations 1981-82*. Cumberland and Westmoorland Antiquarian and Archaeological Society Research Series 5, Kendal.

McGlade J (1995) "Archaeology and the ecodynamics of human-modified landescapes." *Antiquity* **69**: 113-132.

McGlade J & McGlade JM (1989) Modelling the innovative component of social change. In: van der Leeuw SE & Torrence R *What's New? A Closer Look at the Process of Innovation*. Boston, Unwin Hyman Ltd: 281-299.

Meadows KI (1994) "You Are What You Eat: Diet, Identity and Romanisation." *TRAC 94. Proceedings of the Fourth Annual Theoretical Roman Archaeology Conference*: 133-140.

Merton R (1968) *Social Theory and Social Structure*, New York, Free Press.

Metcalfe JS (1986) Technological Innovation and the Competitive Process. In: Hall P *Technology, Innovation and Economic Policy*. Southampton, Philip Allan.

Middleton P (1979) Army supply in Gaul: An hypothesis for Roman Britain. *Invasion and Response*. J. H. Burnham BC &. Oxford, B.A.R.: 81-97.

Midgley G, Munlo I, Brown M (1998) "The theory and practice of boundary critique: developing housing services for older people." *Journal of Operational Research Society* **49**: 467-478.

Millett M (1990) *The Romanization of Britain. An essay in archaeological interpretation*. Cambridge, Cambridge University Press.

Moldovan S & Goldenberg J (2004). "Cellular automata modelling of resistance to innovations: Effects and solutions." *Technological Forecasting and Social Change* **71**: 425-442.

Nash D (1976) The Growth of Urban Society in France. In: Cunliffe B & Rowley T *Oppida: the Beginnings of Urbanisation in Barbarian Europe*. Oxford, BAR. **11:** 95-133.

Oelschlägel C (2006) Nahrungswirtschaft, Tierhaltung und Kultpraxis im Moselgebiet in der römischen Kaiserzeit. In: May EU & Benecke N *Beiträge zur Archäozoologie und Prähistorischen Anthropologie* 5. Konstanz.

Ogburn WF (1966). *Social Change*. New York, Viking Press.

Orton C (1992) Quantitative Methods in the 1990s. In: Lock G & Moffett J *CAA91. Computer Applications and Quantitative Methods in Archaeology*. Oxford, B.A.R. **577:** 137-140.

Orton C & Tyers P (1989) Error structures of ceramic assemblages. In: Rahtz S & Richards J *CAA89. Computer Applications and Quantitative Methods in Archaeology*. Oxford, B.A.R. **548:** 275-286.

Ottaway BS (2001) "Innovation, Production and Specialization in Early Prehistoric Copper Metallurgy." *European Journal of Archaeology* **4**(1): 87-112.

Parsons T (1937) *Structure of Social Action*. New York, McGraw Hill.

Patel P & Pavitt K (1995) Patterns of Technological Activity: their Measurement and Interpretation. In: Stoneman P *Handbook of the Economics of Innovation and Technological Change*. Oxford, Cambridge, Blackwell Publishers Inc.: 14-51.

Payne S (1972a) Partial recovery and sample bias: the results of some sieving experiments. In: Higgs ES *Papers in Economic Prehistory*. Cambridge, University Press: 49-64.

Payne S (1972b) On the interpretation of bone samples from archaeological sites. In: Higgs ES *Papers in Economic History*. Cambridge, University Press: 65-81.

Peacock DPS (1978) The Rhine and the problem of Gaulish wine in Roman Britain. In: du Plat Taylor J & Cleere H *Roman shipping and trade: Britain and the Rhine provinces*. London, CBA report. **24:** 49-51.

Peacock DPS & Williams DF (1986) *Amphorae and the Roman economy. An introductory guide*. London, Longman.

Perez C (1983) "Structural change and assimilation of new technologies in the economic and social systems." *Futures* **15**(5): 357-375.

Peters J (1998) *Römische Tierhaltung und Tierzucht. Eine Synthese aus archäozoologischer Untersuchung und schriftlich-bildlicher Überlieferung*. Rahden, Marie Leidorf GmbH.

Pfaffenberger B (1992) "Social Anthropology of Technology." *Annual Review of Anthropology* **21**: 491-516.

Phillips EJ (1977) *Corpus Signorum Imperii Romani. Great Britain*. Oxford, Oxford University Press.

Pitts M (2004) "Regional identities and the social use of ceramics." *TRAC 04. Proceedings of the Fourteenth Annual Theoretical Roman Archaeology Conference*: 50-64.

Pitts M. & Perring D (2006) "The making of Britain's first urban landscapes: The case of Late Iron Age and Roman Essex". *Britannia* XXXVII: 189-212.

Platzer W (1991) *Taschenatlas der Anatomie. Band 1: Bewegungsapparat*. Stuttgart, Thieme.

Polanyi M (1998) *Personal Knowlegde. Towards a Post Critical Philosophy*. London, Routledge.

Pumain D & Favaro JM (2006) A Multi-scalar time geography for spatial modelling. *Time-Geographical approaches to Emergence and Sustanaible Societies (TiGrESS). Final Report. http://www.tigress.ac/reports/final/mstg.pdf*, accessed october 2006.

Raftery B (1972) Irish Hill-Forts, CBA Report No 9: 37-58.

Renfrew C (1972) *The Emergence of Civilisation. The Cyclades and the Aegean in the Third Millennium BC*. London, Methuen & Co.

Renfrew C (1986) Varna and the emergence of wealth in prehistoric Europe. In: Appadurai A *The Social Life of Things: Commodities in Cultural Perspective*. Cambridge, Cambridge University Press: 141-168.

Renfrew C (2005) Characterisation and Exchange Theory. In: Renfrew C & Bahn P. London, *Archaeology. The Key Concepts*. New York, Routledge: 31-35.

Renfrew C & Bahn P (1996) *Archaeology. Theories, Methods and Practice*. London, Thames and Hudson Ltd.

Rielly K (1988). The animal bone. In: Trow SD *Excavations at Ditches Hillfort, North Cerney, Gloucestershire 1982-3*. Transactions of the Bristol and Gloucestershire Archaeological Society: 77-79.

Rielly K & Trow SD (1988) The Ditches animal bone assemblage in its regional context. In: Trow SD *Excavations at Ditches hillfort, North Cerney, Gloucestershire, 1982-3*: 79-82.

Ringrose TJ (1993) Bone counts and statistics: A Critique. *Journal of Archaeological Science* **20**:121-157.

Rittel H & Webber M (1973) "Dilemmas in a General Theory of Planning." *Policy Science* **4**: 155-169.

Rogers EM & Shoemaker FF (1971) *Communication of Innovations*. New York, The Free Press.

Rosenhead J & Mingers J (2001) A New Paradigm of Analysis. In: Rosenhead J & Mingers J *Rational Analysis for a Problematic World Revisited. Problem Structuring Methods for Complexity, Uncertainty and Conflict*. Chichester, John Wiley & Son: 1-19.

Ryder ML (1983) *Sheep & Man*. London, Duckwoth & Co. Ltd.

Sabloff J (2005) Processual Archaeology. In: Renfrew C & Bahn P *Archaeology. The Key Concepts*. Oxford, Routledge: 212-219.

Sahlins M (1974) *Stone Age Economics*. London, Tavistock Publications.

Sanders L (1999) "Modelling within a self-organising or a microsimulation framework: opposite or complementary approaches?" *Cybergeo* **90**.

Sanders L, Favaro JM, Benoît G, Mathian H, Pumain D, Bretagnolle A, Buxeda C, Didelon C, Durand-Dastès, Paulus F, Rozenblat C & Vachiani C (2006) *Dynamics of the European Urban Network. TiGrESS Final Report.*

http://tigress.ac/reports/final/eurosim. pdf, accessed october 2006.

Schecter D (1991) Critical System Thinking in the 1980s: A Connective Summary. In: Flood RL & Jackson MC *Critical Systems Thinking*. Cambridge, Wiley: 213-226.

Schimanck U (2000) *Handeln und Strukturen. Einführung in die akteurstheoretische Soziologie*. Weinheim, Juventa Verlag.

Schulze PM (1990) *Beschreibende Statistik*. München, Oldenbourg.

Schumpeter JA (1939) *Business Cycles. A Theoretical, Historical, and Statistical Analysis of the Capitalist Process*. New York, McGraw Hill Book Company.

Schumpeter JA (1949) *The Theory of Economic Development*. Cambridge, Harvard University Press.

Searle JR (1998) *Mind, language and society. Philosophy in the real world*. New York, Basic Books.

Seaton RAF & Cordey-Hayes M (1993) "The development and application of interactive models of industrial technology transfer." *technovation* **13**(1): 45-53.

Seeman M (1987) Faunal Remains of Excavations during 1978 and 1979. In: Brandt RW, van Groenman-Waateringe W & van der Leeuw SE *Assendelver Polder Papers 1*. Amsterdam, Universitaet van Amsterdam: 91-98.

Shapiro SH (1982) "Collapsing Contingency Tables – A Geometric Approach." *The American Statistician* 36(1): 43-46.

Sharples NM (1991a) Maiden Castle. Excavations and field survey 1985-6. London, English Heritage.

Sharples NM (1991b) *English Heritage Book of Maiden Castle*. London, B.T. Batsford/English Heritage.

Shennan S (2001) "Demography and Cultural Innovation: a Model and its Implications for the Emergence of Modern Human Culture." *Cambridge Archaeological Journal* **11**(1): 5-16.

Sherratt A (1981) Plough and pastoralism: aspects of secondary products revolution. In: Hodder I, Issac G & Hammond N *Pattern of the Past. Studies in honour of David Clarke*. Cambridge, Cambridge University Press: 261-305.

Sherratt A (1996) "Why Wessex? The Avon Route and River Transport in later British Prehistory." *Oxford Journal of Archaeology* **15**(2): 211-235.

Shotwell JA (1955) "An approach to the palaeoecology of mammals." *Ecology* **36**: 327-337.

Simmel G (1908) *Soziologie. Untersuchung über die Formen der Vergesellschaftung*. Berlin, Duncher + Humbolt.

Simon HG (1982a) Eroberung und Verzicht. Die römische Politik in Germanien zwischen 12 v. Chr. und 16 n. Chr. In: Baatz D & Herrmann FR *Die Römer in Hessen*. Stuttgart, Konrad Theiss Verlag: 38-57.

Simon HG (1982b) Die Zeit der Defensive: Die römische Grenzpolitik zwischen 16 und 69 n. Chr. In: Baatz D & Herrmann FR *Die Römer in Hessen*. Stuttgart, Theis: 58-65.

Smith K (2005) Measuring Innovation. In: Fagerberg J, Mowery DC & Nelson RR *The Oxford Handbook of Innovation*. Oxford, Oxford University Press: 148-177.

Speth JD (1983) *Bison Kills and Bone Counts. Decision Making by Ancient Hunters*. Chicago, University of Chicago Press.

Stoneman P (1991) Technological Diffusion: the Viewpoint of Economic Theory. In: Mathias P & Davis JA *Innovation and Technology in Europe. From the Eighteenth Century to the Present Day*. Oxford, Blackwell: 162-184.

Stoneman P (1995) Introduction. In: Stoneman P *Handbook of the Economics of Innovation and Technological Change*. Cambridge, Blackwell Publishers: 1-13.

Sundbo J (1998) *The Theory of Innovation. Entrepreneurs, Technology and Strategy*. Cheltenham, Edward Elgar Publishing.

Swearengen JC & Woodhouse EJ (2001) "Cultural Risk of Technological Innovation: The Case of School Violence." *Technology and Society Magazine* **20**(1): 15-28.

Thurston TL (1999) "The knowable, the doable and the undiscussed: tradition, submission, and the 'becoming' of rural landscapes in Denmark's Iron Age." *Antiquity* **73**: 661-71.

Toit SHC, Steyn AGW & RH Stumpf (1986) *Graphical Exploratory Data Analysis*. New York, Springer.

Tomes A (2003) "UK government science policy: the 'enterprise deficit' fallacy." *technovation* **23**: 785-792.

Trigger BG (1997) *A History of Archaeological Thought*. Cambridge, Cambridge University Press.

Trott P (2002) *Innovation Management and New Product Development*. London, Prentice Hall.

Trow SD (1988) "Excavations at Ditches hillfort, North Cerney, Gloucestershire, 1982-3." *Transactions of the Bristol and Gloucestershire Archaeological Society* **106**: 19-85.

Trow SD (1990) By the northern shores of Ocean. Some observations on acculturation process at the edge of the Roman world. In: Blagg T & Millett M *The Early Roman Empire in the West*. Oxford, Oxbow Books: 103-118.

Tukey JW (1977). *Exploratory Data Analysis*. Reading, Addison-Wesley Publishing.

Tyers P (1996) *Roman Pottery in Britain*. London, B T Batsford Ltd.

van de Ven AH (1986) "Central problems in the management of innovation." *Management Science* **32**(5): 590-607.

van der Leeuw S (1989) Risk, Perception, Innovation. In: van der Leeuw S *What's new? A closer look at the process of innovation*. London, Unwin Hyman: 300-329.

van der Leuw S (1993) Giving the potter a choice. Conceptual aspects of pottery techniques. In: Lemonnier P *Technological Choices. Transformation in material cultures since the Neolithic*. London, New York, Routledge: 238-288.

van der Leeuw S, Audouze F, Argant J, Ballais JL, Beeching A, Bel V, Berger JF, Bois M, Brochier JL, Chouquer G, Favory F, Fiches JL, Gazenbeek M, Girardot JJ, Jung C, Magnin F, Meffre JC, Odiot T, Raynaund C, Thiébault s, Tourneux FP, Tschanz X, Verhagen P & Zannier MP (1998). Land use, settlement pattern and degradation in the Ancienct Rhone Valley. *The Archaeomedes project. Understanding the natural and anthropogenic causes of land degradation and desertification in the Mediterranean basin*. van der Leeuw S. Belgium, EU. **Synthesis:** 175-221.

van der Leeuw S & R Torrence (1989) Introduction: What's new about innovation? In: van der Leeuw S *What's new? A closer look at the process of innovation*. London, Unwin Hyman: 1-15.

van Groenman-Waateringe W (1980) Urbanization and the North-West Frontier of the Roman Empire. In: Hanson WS & Keppie LFJ *Roman Frontier Studies 1979. Papers presented to the 12th International Congress of Roman Frontier Studies*. Oxford, BAR International Series. **3:** 1037-1044.

van Gundy AB (1987) *Creative Problem Solving. A Guide for Trainers and Managment*. New York, Quorom Books.

Vickers G (1968) *Value Systems and Social Process*. London, Tavistock Publications.

Vickers G (1972) *Freedom in a Rocking Boat: Changing values in an unstable society*. London, Penguin Books.

Vickers G (1983) *Human systems are different*. London, Harper & Row.

Vickers G (1995) *The Art of Judgment. A Study of Policy Making. Centenary Edition*. London, SAGE Publications.

von Pierer H (2002) Developing and Marketing an Innovative Idea. Chasing the Goal of Commercial Success. In: von Pierer H & von Oetinger B *A Passion for Ideas. How innovators create the new and shape the world*. West Lafayette, Purdue University: 173-188.

Wainwright GJ & Davies SM (1995) Balksbury Camp Hampshire. Excavations 1973 and 1981. London, English Heritage Archaeological Report 4.

Watzlawick P, Weakland J & Fisch R (1974) *Change. Principles of Problem Formation and Problem Resolution*. New York, W. W. Norton & Company. Inc.

Weber M (1978) *Economy and Society. An outline of interpretive sociology*. Berkeley, Los Angeles, University of California Press.

West MA & Farr JL (1990) Innovation at Work. In: West MA & Farr JL *Innovation and Creativity at Work: Psychological and Organizational Strategies*. New York, Wiley & Sons: 3-13.

Wightman EM (1977) "Military arrangements, native settlements and related developments in early Roman Gaul." *Helinium* **17**: 105-126.

Wightman E (1985) *Gallia Belgica*. London, B.T. Batsford.

Willems H (1997) *Rahmen und Habitus*. Frankfurt, Suhrkamp.

Wilson EO (1975) Sociobiology: *The New Synthesis*. Cambridge, Harvard University Press.

Williams DF & Peacock DPS (1983) The importation of olive-oil into Iron Age and Roman Britain. In: Blazquez Martinez JM & Remesal Rodriquez J *Produccion y comercio del aceite en la antigüedad.* Universidad Complutense Madrid: 263-280.

Winder N (1996) "When is a time-trend not a time-trend? Scale and Profile at Bronze Age Phylakopi (Melos)." *Circaea* **11**(2): 49-64.

Winder N (1997) The Impact of Recovery Methods on Assemblage Variation: The Confounding Problem. In: Bailey G *Klithi: Paleolithic settlement and Quaternar environments in Northwest Greece: Volume 1: Excavation and Intra-Site Analysis.* Cambridge, McDonald Institute for Archaeological Research. **1:** 105-113.

Winder N (2004) "Towards a Theory of Knowledge Systems for Integrative Socio-Natural Science." *Research in Human Ecology* **11**(2): 118-132.

Winder N (2005) *Breaking the Phoenix Cycle: an integrative approach to Innovation and Cultural Ecodynamics. http://www.aquadapt.net*, accessed October 2006.

Winder N (2006) *Space, Time and Innovation: the Time Geography of Subsidiarity. TiGrESS Final Report. http://tigress.ac/reports/final/WP1_innovation.pdf*, accessed October 2006.

Woodward PJ (1987) The Excavation of an Iron Age and Romano-British Settlement at Rope Lake Hole, Corfe Castle, Dorset, Dorset Natural History and Archaeological Society Monograph 6: 125-184.

Woolf G (1993) "Rethinking the Oppida." *Oxford Journal of Archaeology* **12**(2): 223-234.

Woolf G (1998) *Becoming Roman. The Origins of Provincial Civilazation in Gaul.* Cambridge, Cambridge University Press.

Ziman J (2000) Evolutionary models for technological change. In: Ziman J *Technological Innovation as an Evolutionary Process.* Cambridge, Cambridge University Press: 3-12.

APPENDIX 1

Raw data for the 46 British Iron Age sites published in Hambleton 1999, appendix 3. Frequencies for cow, pig and sheep elements are given as NISP numbers. The site reports for Danebury and Mount Batten gave 2 frequencies for most bones for the distal and proximal end. To adjust them to the other site reports the higher value is chosen to represent the bone.

For Eldon Seat, Hartigans, Heathrow EIA and Heathrow LIA-RB no pig frequencies were given, while Baldock LIA was lacking sheep frequencies. Consequently these 5 sites had to be excluded from some of the exploratory data analyses.

*: missing value
SUM NISP: frequencies of bones for which both species and skeletal element could be identified
TOTAL NISP: frequencies of all bones that could be allocated to one of the species (Hambleton 1999: appendix 2). Where time periods did not match between the frequencies of skeletal elements given in appendix 3 (Hambleton 1999) and the NISP numbers in appendix 2, the sites were excluded from TOTAL NISP comparisons.

CATTLE

Site Name	Abbotstone Down	Baldock LIA	Balksbury EIA	Balksbury MIA	Bancroft LBA-EIA	Bancroft LIA-ERB
c_atlas axis	*	*	*	*	*	*
c_astragalus	5	10	1	15	*	*
c_calcaneum	9	10	2	22	*	*
c_femur	9	7	9	83	0	7
c_fem_distal	*	*	*	*	*	*
c_fem_proximal	*	*	*	*	*	*
c_humerus	20	13	20	137	4	12
c_hum_distal	*	*	*	*	*	*
c_hum_proximal	*	*	*	*	*	*
c_mandible	43	22	34	217	2	39
c_metacarpal	13	16	10	70	*	*
c_metacarpal_d	*	*	*	*	*	*
c_metacarpal_p	*	*	*	*	*	*
c_metapodials	*	*	*	*	3	31
c_metatarsal	18	12	12	63	*	*
c_metatarsal_d	*	*	*	*	*	*
c_metatarsal_p	*	*	*	*	*	*
c_pelvis	12	12	20	94	0	11
c_phal1	7	*	3	14	*	*
c_phal2	2	*	0	3	*	*
c_phal3	1	*	0	2	*	*
c_phalanges	*	*	*	*	0	5
c_radius	16	11	19	75	*	*
c_radius_d	*	*	*	*	*	*
c_radius_p	*	*	*	*	*	*
c_radius ulna	*	*	*	*	4	23
c_rib	0	*	0	6	0	5
c_scapula	29	11	23	123	3	7
c_tibia_d	*	*	*	*	*	*
c_tibia_p	*	*	*	*	*	*
c_tibia	18	7	14	105	1	14
c_ulna	13	3	6	43	*	*
c_ulna_p	*	*	*	*	*	*
c_vertebra	6	*	17	68	3	19
SUM NISP	221	134	190	1140	20	173
TOTAL NISP	326	22	272	1542	-	256

CATTLE

Site Name	Bishopstone MIA-LIA	Bramdean	Brighton Hill South	Burgh LIA	Catcote LIA-RB	Cat's Water total IA
c_atlas axis	*	*	*	*	*	*
c_astragalus	0	3	9	4	18	*
c_calcaneum	0	1	16	8	16	*
c_femur	0	6	37	34	11	137
c_fem_distal	*	*	*	*	*	*
c_fem_proximal	*	*	*	*	*	*
c_humerus	0	2	60	36	34	170
c_hum_distal	*	*	*	*	*	*
c_hum_proximal	*	*	*	*	*	*
c_mandible	13	15	173	107	22	308
c_metacarpal	0	*	28	30	24	181
c_metacarpal_d	*	*	*	*	*	*
c_metacarpal_p	*	*	*	*	*	*
c_metapodials	*	8	*	*	*	*
c_metatarsal	0	*	48	56	23	144
c_metatarsal_d	*	*	*	*	*	*
c_metatarsal_p	*	*	*	*	*	*
c_pelvis	1	4	34	40	6	66
c_phal1	*	0	5	12	36	*
c_phal2	*	2	6	10	22	*
c_phal3	*	1	2	1	8	*
c_phalanges	0	*	*	*	*	78
c_radius	0	4	36	55	21	150
c_radius_d	*	*	*	*	*	*
c_radius_p	*	*	*	*	*	*
c_radius ulna	*	*	*	*	*	*
c_rib	*	0	4	291	*	*
c_scapula	1	7	86	55	19	126
c_tibia_d	*	*	*	*	*	*
c_tibia_p	*	*	*	*	*	*
c_tibia	0	5	55	52	23	164
c_ulna	1	4	27	18	10	76
c_ulna_p	*	*	*	*	*	*
c_vertebra	2	8	70	67	28	85
SUM NISP	18	70	696	876	321	1685
TOTAL NISP	304	-	-	585	340	2596

CATTLE

Site Name	Chilbolton Down EIA-MIA	Dalton Parlours MIA-LIA	Danebury EIA	Danebury LIA	Danebury MIA	Danebury MIA-LIA
c_atlas axis	*	*	58	103	19	155
c_astragalus	*	*	60	96	41	159
c_calcaneum	3	*	36	92	38	135
c_femur	8	6	*	*	*	*
c_fem_distal	*	*	39	46	17	113
c_fem_proximal	*	*	52	65	18	131
c_humerus	11	5	*	*	*	*
c_hum_distal	*	*	78	136	37	200
c_hum_proximal	*	*	43	47	20	78
c_mandible	8	13	77	156	39	145
c_metacarpal	5	*	*	*	*	*
c_metacarpal_d	*	*	45	69	22	112
c_metacarpal_p	*	*	66	76	26	170
c_metapodials	*	17	*	*	*	*
c_metatarsal	5	*	*	*	*	*
c_metatarsal_d	*	*	45	75	23	86
c_metatarsal_p	*	*	64	90	26	132
c_pelvis	6	14	81	123	35	169
c_phal1	2	*	147	159	62	272
c_phal2	3	*	115	113	48	179
c_phal3	0	*	79	65	30	132
c_phalanges	*	42	*	*	*	*
c_radius	7	8	*	*	*	*
c_radius_d	*	*	43	66	27	123
c_radius_p	*	*	68	142	45	216
c_radius ulna	*	*	*	*	*	*
c_rib	*	*	*	*	*	*
c_scapula	9	11	71	141	46	216
c_tibia_d	*	*	54	114	40	158
c_tibia_p	*	*	40	53	20	98
c_tibia	3	7	*	*	*	*
c_ulna	2	3	*	*	*	*
c_ulna_p	*	*	60	119	31	169
c_vertebra	9	*	*	*	*	*
SUM NISP	81	126	1166†	1790†	581†	2738†
TOTAL NISP	113	166	3679	5953	1911	8355

† For all Danebury sites, rather than adding all frequencies, the SUM NISP values were calculated without the lower frequencies for bones for which values for both distal and proximal ends were given

CATTLE

Site Name	Ditches LIA-ERB	Dragonby MIA-LIA	Edix Hill LIA	Eldon Seat	Farningham Hill LIA	Grimthorpe IA
c_atlas axis	*	*	*	*	*	*
c_astragalus	22	21	9	15	5	9
c_calcaneum	30	36	15	11	1	8
c_femur	92	86	1	12	13	11
c_fem_distal	*	*	*	*	*	*
c_fem_proximal	*	*	*	*	*	*
c_humerus	68	78	6	13	19	17
c_hum_distal	*	*	*	*	*	*
c_hum_proximal	*	*	*	*	*	*
c_mandible	207	150	*	95	41	31
c_metacarpal	36	69	4	61	24	*
c_metacarpal_d	*	*	*	*	*	*
c_metacarpal_p	*	*	*	*	*	*
c_metapodials	*	*	*	*	*	53
c_metatarsal	74	90	4	24	4	*
c_metatarsal_d	*	*	*	*	*	*
c_metatarsal_p	*	*	*	*	*	*
c_pelvis	104	103	6	*	23	22
c_phal1	64	90	12	33	*	*
c_phal2	42	45	*	13	*	*
c_phal3	44	29	4	9	*	*
c_phalanges	*	*	*	*	15	29
c_radius	47	*	3	20	14	*
c_radius_d	*	*	*	*	*	*
c_radius_p	*	*	*	*	*	*
c_radius ulna	*	186	*	*	*	30
c_rib	*	*	*	*	4	15
c_scapula	188	124	8	*	30	13
c_tibia_d	*	*	*	*	*	*
c_tibia_p	*	*	*	*	*	*
c_tibia	94	104	12	22	16	19
c_ulna	52	*	*	10	7	*
c_ulna_p	*	*	*	*	*	*
c_vertebra	18	*	*	*	23	21
SUM NISP	1182	1211	84	338	239	278
TOTAL NISP	2028	2745	177	-	530	403

CATTLE

Site Name	Hartigans	Heathrow EIA	Heathrow LIA-RB	Maiden Castle EIA-LIA	Market Deeping MIA-LIA	Meare 1984 LIA
c_atlas axis	*	*	*	*	*	*
c_astragalus	2	1	3	11	6	9
c_calcaneum	1	1	1	17	7	5
c_femur	5	1	0	41	1	18
c_fem_distal	*	*	*	*	*	*
c_fem_proximal	*	*	*	*	*	*
c_humerus	2	3	2	45	7	17
c_hum_distal	*	*	*	*	*	*
c_hum_proximal	*	*	*	*	*	*
c_mandible	42	11	8	71	*	49
c_metacarpal	6	3	2	48	4	11
c_metacarpal_d	*	*	*	*	*	*
c_metacarpal_p	*	*	*	*	*	*
c_metapodials	*	*	*	*	*	*
c_metatarsal	7	3	1	52	3	9
c_metatarsal_d	*	*	*	*	*	*
c_metatarsal_p	*	*	*	*	*	*
c_pelvis	9	6	2	35	3	5
c_phal1	4	*	*	43	9	*
c_phal2	1	*	*	29	*	*
c_phal3	0	*	*	13	*	*
c_phalanges	*	*	*	*	*	7
c_radius	9	6	3	59	*	23
c_radius_d	*	*	*	*	*	*
c_radius_p	*	*	*	*	*	*
c_radius ulna	*	*	*	*	9	*
c_rib	*	*	*	116	*	*
c_scapula	1	5	2	43	3	8
c_tibia_d	*	*	*	*	*	*
c_tibia_p	*	*	*	*	*	*
c_tibia	5	3	1	41	8	23
c_ulna	3	2	0	37	*	12
c_ulna_p	*	*	*	*	*	*
c_vertebra	3	*	*	79	*	7
SUM NISP	100	45	25	780	60	203
TOTAL NISP	144	-	-	950	131	419

CATTLE

Site Name	Meare West 1979 MIA-LIA	Mount Batten total IA	Old Down Farm	Ower LIA-RB	Owslebury	Pennyland MIA
c_atlas axis	*	20	*	*	*	*
c_astragalus	11	9	17	*	76	*
c_calcaneum	14	15	15	*	91	*
c_femur	15	*	36	3	244	38
c_fem_distal	*	5	*	*	*	*
c_fem_proximal	*	4	*	*	*	*
c_humerus	11	*	68	4	306	43
c_hum_distal	*	4	*	*	*	*
c_hum_proximal	*	2	*	*	*	*
c_mandible	*	12	75	12	1084	68
c_metacarpal	10	*	32	*	235	*
c_metacarpal_d	*	5	*	*	*	*
c_metacarpal_p	*	19	*	*	*	*
c_metapodials	*	*	*	9	*	64
c_metatarsal	9	*	30	*	275	*
c_metatarsal_d	*	4	*	*	*	*
c_metatarsal_p	*	20	*	*	*	*
c_pelvis	*	13	73	8	253	40
c_phal1	*	27	*	*	141	*
c_phal2	*	34	*	*	67	*
c_phal3	*	21	*	*	28	*
c_phalanges	*	*	64	5	*	37
c_radius	7	*	49	5	236	*
c_radius_d	*	3	*	*	*	*
c_radius_p	*	6	*	*	*	*
c_radius ulna	*	*	*	*	*	52
c_rib	*	*	17	5	27	13
c_scapula	13	17	38	0	444	31
c_tibia_d	*	9	*	*	*	*
c_tibia_p	*	0	*	*	*	*
c_tibia	6	*	48	4	285	57
c_ulna	*	*	30	2	173	*
c_ulna_p	*	7	*	*	*	*
c_vertebra	*	*	131	6	397	104
SUM NISP	96	238†	723	63	4362	547
TOTAL NISP	442	1423	-	101	-	710

† For Mount Batten, rather than adding all frequencies, the SUM NISP values were calculated without the lower frequencies for bones for which values for both distal and proximal ends were given

CATTLE

Site Name	Port Seton LIA-ERB	Poundbury total IA	Puckeridge-Braughing LIA-RB	Rope Lake Hole IA-RB	Skeleton Green LIA-RB	Uley Bury
c_atlas axis	*	*	*	*	*	*
c_astragalus	13	19	37	*	*	0
c_calcaneum	7	7	57	*	*	0
c_femur	14	8	63	21	46	8
c_fem_distal	*	*	*	*	*	*
c_fem_proximal	*	*	*	*	*	*
c_humerus	23	15	117	40	62	19
c_hum_distal	*	*	*	*	*	*
c_hum_proximal	*	*	*	*	*	*
c_mandible	89	18	88	103	42	64
c_metacarpal	27	22	73	*	17	*
c_metacarpal_d	*	*	*	*	*	*
c_metacarpal_p	*	*	*	*	*	*
c_metapodials	*	*	*	98	*	18
c_metatarsal	37	11	77	*	34	*
c_metatarsal_d	*	*	*	*	*	*
c_metatarsal_p	*	*	*	*	*	*
c_pelvis	36	7	93	40	41	10
c_phal1	11	*	83	*	*	*
c_phal2	4	*	25	*	*	*
c_phal3	0	*	23	*	*	*
c_phalanges	*	11	*	62	49	29
c_radius	24	14	111	29	49	*
c_radius_d	*	*	*	*	*	*
c_radius_p	*	*	*	*	*	*
c_radius ulna	*	*	*	*	*	19
c_rib	1	38	*	80	*	51
c_scapula	30	23	139	46	77	19
c_tibia_d	*	*	*	*	*	*
c_tibia_p	*	*	*	*	*	*
c_tibia	22	15	132	25	64	21
c_ulna	11	7	31	22	24	*
c_ulna_p	*	*	*	*	*	*
c_vertebra	25	40	6	43	125	39
SUM NISP	374	255	1155	609	630	297
TOTAL NISP	536	432	1348	-	786	-

CATTLE

Site Name	Wakerley MIA-LIA	Wavendon Gate	West Stow MIA-LIA	Winklebury EIA-MIA
c_atlas axis	*	*	*	*
c_astragalus	12	3	31	8
c_calcaneum	7	11	34	12
c_femur	5	40	51	28
c_fem_distal	*	*	*	*
c_fem_proximal	*	*	*	*
c_humerus	15	25	54	20
c_hum_distal	*	*	*	*
c_hum_proximal	*	*	*	*
c_mandible	46	74	127	69
c_metacarpal	20	9	61	17
c_metacarpal_d	*	*	*	*
c_metacarpal_p	*	*	*	*
c_metapodials	*	*	*	*
c_metatarsal	18	7	62	27
c_metatarsal_d	*	*	*	*
c_metatarsal_p	*	*	*	*
c_pelvis	*	23	38	59
c_phal1	15	4	53	18
c_phal2	5	0	22	15
c_phal3	8	0	18	12
c_phalanges	*	*	*	*
c_radius	19	32	59	26
c_radius_d	*	*	*	*
c_radius_p	*	*	*	*
c_radius ulna	*	*	*	*
c_rib	*	*	*	174
c_scapula	10	32	69	47
c_tibia_d	*	*	*	*
c_tibia_p	*	*	*	*
c_tibia	16	36	76	36
c_ulna	0	10	24	13
c_ulna_p	*	*	*	*
c_vertebra	*	6	20	167
SUM NISP	196	312	799	748
TOTAL NISP	196	413	1390	752

PIG

Site Name	Abbotstone Down	Baldock LIA	Balksbury EIA	Balksbury MIA	Bancroft LBA-EIA	Bancroft LIA-ERB
p_astragalus	0	2	0	0	*	*
p_axis atlas	*	*	*	*	*	*
p_calcaneum	0	2	1	3	*	*
p_femur	4	2	5	0	0	1
p_femur_d	*	*	*	*	*	*
p_femur_p	*	*	*	*	*	*
p_humerus	3	11	5	28	1	9
p_humerus_d	*	*	*	*	*	*
p_humerus_p	*	*	*	*	*	*
p_mandible	19	14	10	46	9	13
p_metacarpal	*	0	1	2	*	*
p_metacarpal_d	*	*	*	*	*	*
p_metacarpal_p	*	*	*	*	*	*
p_metapodials	1	*	*	*	2	3
p_metatarsals	*	0	5	3	*	*
p_metatarsal_d	*	*	*	*	*	*
p_metatarsal_p	*	*	*	*	*	*
p_pelvis	2	4	4	8	1	3
p_phal1	0	*	1	0	*	*
p_phal2	0	*	0	0	*	*
p_phal3	0	*	0	0	*	*
p_phalanges	*	*	*	*	2	2
p_radius	1	4	2	12	*	*
p_radius_d	*	*	*	*	*	*
p_radius_p	*	*	*	*	*	*
p_radius ulna	*	*	*	*	2	4
p_ribs	1	*	1	2	0	0
p_scapula	3	9	4	30	0	10
p_tibia	6	3	7	20	1	5
p_tibia_d	*	*	*	*	*	*
p_tibia_p	*	*	*	*	*	*
p_ulna	4	6	1	9	*	*
p_ulna_p	*	*	*	*	*	*
p_vertebra	2	*	2	7	0	0
SUM NISP	46	57	49	170	18	50
TOTAL NISP	63	-	72	282	-	71

PIG

Site Name	Bishopstone MIA-LIA	Bramdean	Brighton Hill South	Burgh LIA	Catcote LIA-RB	Cat's Water total IA
p_astragalus	0	0	1	0	0	*
p_axis atlas	*	*	*	*	*	*
p_calcaneum	0	0	1	0	0	*
p_femur	0	1	14	13	1	24
p_femur_d	*	*	*	*	*	*
p_femur_p	*	*	*	*	*	*
p_humerus	0	5	13	14	3	26
p_humerus_d	*	*	*	*	*	*
p_humerus_p	*	*	*	*	*	*
p_mandible	13	16	50	44	17	57
p_metacarpal	0	*	1	7	*	*
p_metacarpal_d	*	*	*	*	*	*
p_metacarpal_p	*	*	*	*	*	*
p_metapodials	*	4	*	*	8	21
p_metatarsals	0	*	2	9	*	*
p_metatarsal_d	*	*	*	*	*	*
p_metatarsal_p	*	*	*	*	*	*
p_pelvis	1	1	7	24	2	20
p_phal1	*	4	3	3	0	*
p_phal2	*	0	4	0	0	*
p_phal3	*	3	1	0	0	*
p_phalanges	0	*	*	*	*	1
p_radius	0	3	11	6	0	10
p_radius_d	*	*	*	*	*	*
p_radius_p	*	*	*	*	*	*
p_radius ulna	*	*	*	*	*	*
p_ribs	*	0	2	*	*	*
p_scapula	1	8	27	11	4	39
p_tibia	0	2	13	16	1	42
p_tibia_d	*	*	*	*	*	*
p_tibia_p	*	*	*	*	*	*
p_ulna	1	1	11	17	0	22
p_ulna_p	*	*	*	*	*	*
p_vertebra	2	4	10	2	0	2
SUM NISP	18	52	171	166	36	264
TOTAL NISP	78	-	-	178	47	393

PIG

Site Name	Chilbolton Down EIA-MIA	Dalton Parlours MIA-LIA	Danebury EIA	Danebury LIA	Danebury MIA	Danebury MIA-LIA
p_astragalus	*	*	37	38	29	95
p_axis atlas	*	*	43	89	20	89
p_calcaneum	*	*	52	44	26	95
p_femur	2	2	*	*	*	*
p_femur_d	*	*	31	21	31	79
p_femur_p	*	*	36	13	25	70
p_humerus	1	2	*	*	*	*
p_humerus_d	*	*	42	74	42	145
p_humerus_p	*	*	27	22	22	60
p_mandible	1	3	80	149	67	263
p_metacarpal	0	*	*	*	*	*
p_metacarpal_d	*	*	48	39	38	107
p_metacarpal_p	*	*	54	52	46	138
p_metapodials	*	1	*	*	*	*
p_metatarsals	0	*	*	*	*	*
p_metatarsal_d	*	*	41	30	25	74
p_metatarsal_p	*	*	46	38	31	101
p_pelvis	0	2	44	62	40	125
p_phal1	0	*	99	83	82	187
p_phal2	0	*	60	58	50	101
p_phal3	1	*	36	20	29	66
p_phalanges	*	6	*	*	*	*
p_radius	0	1	*	*	*	*
p_radius_d	*	*	24	25	27	56
p_radius_p	*	*	47	56	38	116
p_radius ulna	*	*	*	*	*	*
p_ribs	*	*	*	*	*	*
p_scapula	1	0	55	81	37	174
p_tibia	0	1	*	*	*	*
p_tibia_d	*	*	47	66	35	112
p_tibia_p	*	*	34	21	21	53
p_ulna	0	3	*	*	*	*
p_ulna_p	*	*	57	80	44	162
p_vertebra	0	*	*	*	*	*
SUM NISP	6	21	835†	1011†	647†	2048†
TOTAL NISP	15	34	2275	2162	1681	5338

† For all Danebury sites, rather than adding all frequencies, the SUM NISP values were calculated without the lower frequencies for bones for which values for both distal and proximal ends were given

PIG

Site Name	Ditches LIA-ERB	Dragonby MIA-LIA	Edix Hill LIA	Eldon Seat	Farningham Hill LIA	Grimthorpe IA
p_astragalus	10	11	2	*	0	0
p_axis atlas	*	*	*	*	*	*
p_calcaneum	17	18	4	*	0	2
p_femur	16	37	3	*	0	1
p_femur_d	*	*	*	*	*	*
p_femur_p	*	*	*	*	*	*
p_humerus	31	65	6	*	6	8
p_humerus_d	*	*	*	*	*	*
p_humerus_p	*	*	*	*	*	*
p_mandible	90	120	*	*	7	7
p_metacarpal	23	18	4	*	0	*
p_metacarpal_d	*	*	*	*	*	*
p_metacarpal_p	*	*	*	*	*	*
p_metapodials	*	*	*	*	*	1
p_metatarsals	13	15	1	*	0	*
p_metatarsal_d	*	*	*	*	*	*
p_metatarsal_p	*	*	*	*	*	*
p_pelvis	31	21	3	*	6	3
p_phal1	12	16	10	*	*	*
p_phal2	6	11	*	*	*	*
p_phal3	4	1	1	*	*	*
p_phalanges	*	*	*	*	3	1
p_radius	23	*	3	*	0	*
p_radius_d	*	*	*	*	*	*
p_radius_p	*	*	*	*	*	*
p_radius ulna	*	63	*	*	*	12
p_ribs	*	*	*	*	0	0
p_scapula	34	73	4	*	6	3
p_tibia	22	52	2	*	3	1
p_tibia_d	*	*	*	*	*	*
p_tibia_p	*	*	*	*	*	*
p_ulna	20	*	*	*	2	*
p_ulna_p	*	*	*	*	*	*
p_vertebra	32	*	*	*	1	0
SUM NISP	384	521	43	0	34	39
TOTAL NISP	668	102	102	-	65	57

PIG

Site Name	Hartigans	Heathrow EIA	Heathrow LIA-RB	Maiden Castle EIA-LIA	Market Deeping MIA-LIA	Meare 1984 LIA
p_astragalus	*	*	*	2	0	3
p_axis atlas	*	*	*	*	*	*
p_calcaneum	*	*	*	6	0	3
p_femur	*	*	*	8	0	18
p_femur_d	*	*	*	*	*	*
p_femur_p	*	*	*	*	*	*
p_humerus	*	*	*	30	2	14
p_humerus_d	*	*	*	*	*	*
p_humerus_p	*	*	*	*	*	*
p_mandible	*	*	*	36	*	19
p_metacarpal	*	*	*	8	1	6
p_metacarpal_d	*	*	*	*	*	*
p_metacarpal_p	*	*	*	*	*	*
p_metapodials	*	*	*	*	*	*
p_metatarsals	*	*	*	9	1	1
p_metatarsal_d	*	*	*	*	*	*
p_metatarsal_p	*	*	*	*	*	*
p_pelvis	*	*	*	18	3	15
p_phal1	*	*	*	7	*	*
p_phal2	*	*	*	7	*	*
p_phal3	*	*	*	7	*	*
p_phalanges	*	*	*	*	6	24
p_radius	*	*	*	18	*	2
p_radius_d	*	*	*	*	*	*
p_radius_p	*	*	*	*	*	*
p_radius ulna	*	*	*	*	3	*
p_ribs	*	*	*	8	*	*
p_scapula	*	*	*	42	1	9
p_tibia	*	*	*	15	4	17
p_tibia_d	*	*	*	*	*	*
p_tibia_p	*	*	*	*	*	*
p_ulna	*	*	*	24	*	2
p_ulna_p	*	*	*	*	*	*
p_vertebra	*	*	*	41	*	1
SUM NISP	0	0	0	286	21	134
TOTAL NISP	2	-	-	405	22	221

PIG

Site Name	Meare West 1979 MIA-LIA	Mount Batten total IA	Old Down Farm	Ower LIA-RB	Owslebury	Pennyland MIA
p_astragalus	9	4	4	*	37	*
p_axis atlas	*	68	*	*	*	*
p_calcaneum	10	7	6	*	38	*
p_femur	2	*	4	9	143	1
p_femur_d	*	0	*	*	*	*
p_femur_p	*	0	*	*	*	*
p_humerus	2	*	16	4	206	8
p_humerus_d	*	9	*	*	*	*
p_humerus_p	*	2	*	*	*	*
p_mandible	*	72	40	41	685	16
p_metacarpal	6	*	9	*	35	*
p_metacarpal_d	*	2	*	*	*	*
p_metacarpal_p	*	12	*	*	*	*
p_metapodials	*	*	*	3	*	3
p_metatarsals	6	*	6	*	34	*
p_metatarsal_d	*	3	*	*	*	*
p_metatarsal_p	*	12	*	*	*	*
p_pelvis	*	8	4	4	97	3
p_phal1	*	15	*	*	79	*
p_phal2	*	6	*	*	48	*
p_phal3	*	12	*	*	23	*
p_phalanges	*	*	12	4	*	1
p_radius	4	*	4	4	64	*
p_radius_d	*	1	*	*	*	*
p_radius_p	*	9	*	*	*	*
p_radius ulna	*	*	*	*	*	6
p_ribs	*	*	1	1	53	1
p_scapula	2	22	9	7	167	8
p_tibia	6	*	10	8	174	8
p_tibia_d	*	5	*	*	*	*
p_tibia_p	*	2	*	*	*	*
p_ulna	*	*	18	2	93	*
p_ulna_p	*	3	*	*	*	*
p_vertebra	*	*	4	5	120	1
SUM NISP	47	264[†]	147	92	2096	56
TOTAL NISP	238	781	-	238	-	94

[†] For Mount Batten, rather than adding all frequencies, the SUM NISP values were calculated without the lower frequencies for bones for which values for both distal and proximal ends were given

PIG

Site Name	Port Seton LIA-ERB	Poundbury total IA	Puckeridge-Braughing LIA-RB	Rope Lake Hole IA-RB	Skeleton Green LIA-RB	Uley Bury
p_astragalus	0	1	6	*	*	0
p_axis atlas	*	*	*	*	*	*
p_calcaneum	1	0	42	*	*	0
p_femur	0	8	76	2	43	0
p_femur_d	*	*	*	*	*	*
p_femur_p	*	*	*	*	*	*
p_humerus	4	4	85	10	49	5
p_humerus_d	*	*	*	*	*	*
p_humerus_p	*	*	*	*	*	*
p_mandible	25	20	171	30	229	47
p_metacarpal	1	3	55	*	47	*
p_metacarpal_d	*	*	*	*	*	*
p_metacarpal_p	*	*	*	*	*	*
p_metapodials	*	*	*	12	*	3
p_metatarsals	2	1	38	*	45	*
p_metatarsal_d	*	*	*	*	*	*
p_metatarsal_p	*	*	*	*	*	*
p_pelvis	1	5	85	4	75	6
p_phal1	0	*	24	*	*	*
p_phal2	0	*	2	*	*	*
p_phal3	0	*	0	*	*	*
p_phalanges	*	1	*	6	24	2
p_radius	2	5	90	9	42	*
p_radius_d	*	*	*	*	*	*
p_radius_p	*	*	*	*	*	*
p_radius ulna	*	*	*	*	*	6
p_ribs	3	27	*	2	*	0
p_scapula	4	10	103	16	65	23
p_tibia	0	4	140	10	73	7
p_tibia_d	*	*	*	*	*	*
p_tibia_p	*	*	*	*	*	*
p_ulna	2	1	59	7	52	*
p_ulna_p	*	*	*	*	*	*
p_vertebra	1	4	2	1	116	0
SUM NISP	46	94	978	109	860	99
TOTAL NISP	80	120	1412	-	1202	-

PIG

Site Name	Wakerley MIA-LIA	Wavendon Gate	West Stow MIA-LIA	Winklebury EIA-MIA
p_astragalus	0	0	8	12
p_axis atlas	*	*	*	*
p_calcaneum	4	0	3	14
p_femur	0	0	9	8
p_femur_d	*	*	*	*
p_femur_p	*	*	*	*
p_humerus	8	0	28	18
p_humerus_d	*	*	*	*
p_humerus_p	*	*	*	*
p_mandible	*	3	34	40
p_metacarpal	0	0	6	0
p_metacarpal_d	*	*	*	*
p_metacarpal_p	*	*	*	*
p_metapodials	*	*	*	*
p_metatarsals	0	0	6	0
p_metatarsal_d	*	*	*	*
p_metatarsal_p	*	*	*	*
p_pelvis	*	4	15	8
p_phal1	0	*	5	12
p_phal2	0	*	0	6
p_phal3	0	*	0	0
p_phalanges	*	0	*	*
p_radius	8	1	4	28
p_radius_d	*	*	*	*
p_radius_p	*	*	*	*
p_radius ulna	*	*	*	*
p_ribs	*	*	*	1
p_scapula	9	1	26	36
p_tibia	13	0	3	26
p_tibia_d	*	*	*	*
p_tibia_p	*	*	*	*
p_ulna	10	0	6	22
p_ulna_p	*	*	*	*
p_vertebra	*	0	2	32
SUM NISP	52	9	155	263
TOTAL NISP	52	11	270	263

SHEEP

Site Name	Abbotstone Down	Baldock LIA	Balksbury EIA	Balksbury MIA	Bancroft LBA-EIA	Bancroft LIA-ERB
s_astragalus	0	*	0	3	*	*
s_atlas axis	*	*	*	*	*	*
s_calcaneum	1	*	5	14	*	*
s_femur	3	*	11	93	0	3
s_femur_d	*	*	*	*	*	*
s_femur_p	*	*	*	*	*	*
s_humerus	8	*	16	96	2	14
s_humerus_d	*	*	*	*	*	*
s_humerus_p	*	*	*	*	*	*
s_mandible	53	*	39	274	3	27
s_metacarpal	13	*	24	143	*	*
s_metacarpal_d	*	*	*	*	*	*
s_metacarpal_p	*	*	*	*	*	*
s_metapodials	*	*	*	*	5	21
s_metatarsal	29	*	39	225	*	*
s_metatarsal_d	*	*	*	*	*	*
s_metatarsal_p	*	*	*	*	*	*
s_pelvis	2	*	11	80	2	11
s_phal1	5	*	8	18	*	*
s_phal2	0	*	0	0	*	*
s_phal3	1	*	0	0	*	*
s_phalanges	*	*	*	*	1	3
s_radius	21	*	41	208	*	*
s_radius_d	*	*	*	*	*	*
s_radius_p	*	*	*	*	*	*
s_radius ulna	*	*	*	*	6	22
s_ribs	1	*	1	0	0	12
s_scapula	4	*	8	91	0	9
s_tibia	40	*	47	279	0	32
s_tibia_d	*	*	*	*	*	*
s_tibia_p	*	*	*	*	*	*
s_ulna	0	*	13	46	*	*
s_ulna_p	*	*	*	*	*	*
s_vertebra	2	*	7	157	1	9
SUM NISP	183	0	270	1727	20	163
TOTAL NISP	258	-	420	2308	-	208

SHEEP

Site Name	Bishopstone MIA-LIA	Bramdean	Brighton Hill South	Burgh LIA	Catcote LIA-RB	Cat's Water total IA
s_astragalus	1	6	6	2	13	*
s_atlas axis	*	*	*	*	*	*
s_calcaneum	1	2	9	4	17	*
s_femur	2	8	41	38	13	51
s_femur_d	*	*	*	*	*	*
s_femur_p	*	*	*	*	*	*
s_humerus	9	10	58	38	36	118
s_humerus_d	*	*	*	*	*	*
s_humerus_p	*	*	*	*	*	*
s_mandible	11	59	181	140	40	253
s_metacarpal	12	*	41	80	16	154
s_metacarpal_d	*	*	*	*	*	*
s_metacarpal_p	*	*	*	*	*	*
s_metapodials	*	10	*	*	*	*
s_metatarsal	11	*	107	84	24	181
s_metatarsal_d	*	*	*	*	*	*
s_metatarsal_p	*	*	*	*	*	*
s_pelvis	2	3	35	28	12	43
s_phal1	*	2	17	18	0	*
s_phal2	*	3	7	0	0	*
s_phal3	*	1	5	0	0	*
s_phalanges	3	*	*	*	*	25
s_radius	1	7	93	90	27	254
s_radius_d	*	*	*	*	*	*
s_radius_p	*	*	*	*	*	*
s_radius ulna	*	*	*	*	*	*
s_ribs	*	37	7	234	*	*
s_scapula	6	4	30	60	25	67
s_tibia	14	14	177	113	33	333
s_tibia_d	*	*	*	*	*	*
s_tibia_p	*	*	*	*	*	*
s_ulna	2	1	14	8	15	45
s_ulna_p	*	*	*	*	*	*
s_vertebra	2	13	28	42	55	69
SUM NISP	77	180	856	979	326	1593
TOTAL NISP	247	-	-	697	349	2224

SHEEP

Site Name	Chilbolton Down EIA-MIA	Dalton Parlours MIA-LIA	Danebury EIA	Danebury LIA	Danebury MIA	Danebury MIA-LIA
s_astragalus	*	*	179	321	120	362
s_atlas axis	*	*	143	324	73	472
s_calcaneum	0	*	145	292	104	363
s_femur	7	18	*	*	*	*
s_femur_d	*	*	105	244	88	353
s_femur_p	*	*	143	259	95	433
s_humerus	5	8	*	*	*	*
s_humerus_d	*	*	263	588	142	782
s_humerus_p	*	*	93	162	62	246
s_mandible	24	12	383	1033	241	1510
s_metacarpal	9	*	*	*	*	*
s_metacarpal_d	*	*	174	258	93	408
s_metacarpal_p	*	*	202	339	104	536
s_metapodials	*	54	*	*	*	*
s_metatarsal	8	*	*	*	*	*
s_metatarsal_d	*	*	133	225	78	338
s_metatarsal_p	*	*	178	361	115	524
s_pelvis	6	9	200	486	124	614
s_phal1	9	*	447	600	288	751
s_phal2	1	*	172	265	124	285
s_phal3	1	*	109	102	68	158
s_phalanges	*	239	*	*	*	*
s_radius	14	25	*	*	*	*
s_radius_d	*	*	171	333	80	442
s_radius_p	*	*	230	450	114	651
s_radius ulna	*	*	*	*	*	*
s_ribs	*	*	*	*	*	*
s_scapula	9	8	182	441	106	677
s_tibia	17	23	*	*	*	*
s_tibia_d	*	*	246	489	117	626
s_tibia_p	*	*	132	225	60	316
s_ulna	2	6	*	*	*	*
s_ulna_p	*	*	155	308	85	387
s_vertebra	1	*	*	*	*	*
SUM NISP	113	402	3377†	6658†	2020†	9131†
TOTAL NISP	229	495	9253	18290	5730	25892

† For all Danebury sites, rather than adding all frequencies, the SUM NISP values were calculated without the lower frequencies for bones for which values for both distal and proximal ends were given

SHEEP

Site Name	Ditches LIA-ERB	Dragonby MIA-LIA	Edix Hill LIA	Eldon Seat	Farningham Hill LIA	Grimthorpe IA
s_astragalus	21	27	10	6	2	1
s_atlas axis	*	*	*	*	*	*
s_calcaneum	21	37	11	6	1	1
s_femur	66	185	9	18	7	10
s_femur_d	*	*	*	*	*	*
s_femur_p	*	*	*	*	*	*
s_humerus	90	243	16	32	11	5
s_humerus_d	*	*	*	*	*	*
s_humerus_p	*	*	*	*	*	*
s_mandible	206	455	*	89	28	26
s_metacarpal	35	197	9	35	4	*
s_metacarpal_d	*	*	*	*	*	*
s_metacarpal_p	*	*	*	*	*	*
s_metapodials	*	*	*	*	*	12
s_metatarsal	51	228	7	27	7	*
s_metatarsal_d	*	*	*	*	*	*
s_metatarsal_p	*	*	*	*	*	*
s_pelvis	126	161	6	*	8	17
s_phal1	22	54	7	10	*	*
s_phal2	0	21	*	1	*	*
s_phal3	1	4	2	0	*	*
s_phalanges	*	*	*	*	14	2
s_radius	95	*	9	35	23	*
s_radius_d	*	*	*	*	*	*
s_radius_p	*	*	*	*	*	*
s_radius ulna	*	366	*	*	*	12
s_ribs	*	*	*	*	9	11
s_scapula	161	209	10	*	5	0
s_tibia	173	488	15	42	31	13
s_tibia_d	*	*	*	*	*	*
s_tibia_p	*	*	*	*	*	*
s_ulna	36	*	*	11	4	*
s_ulna_p	*	*	*	*	*	*
s_vertebra	36	*	*	*	30	11
SUM NISP	1140	2675	111	312	184	121
TOTAL NISP	1644	5871	337	-	500	184

SHEEP

Site Name	Hartigans	Heathrow EIA	Heathrow LIA-RB	Maiden Castle EIA-LIA	Market Deeping MIA-LIA	Meare 1984 LIA
s_astragalus	0	0	1	32	4	5
s_atlas axis	*	*	*	*	*	*
s_calcaneum	0	0	0	28	1	10
s_femur	0	0	0	171	0	29
s_femur_d	*	*	*	*	*	*
s_femur_p	*	*	*	*	*	*
s_humerus	2	4	1	210	2	37
s_humerus_d	*	*	*	*	*	*
s_humerus_p	*	*	*	*	*	*
s_mandible	5	16	3	286	*	52
s_metacarpal	1	4	0	179	2	39
s_metacarpal_d	*	*	*	*	*	*
s_metacarpal_p	*	*	*	*	*	*
s_metapodials	*	*	*	*	*	*
s_metatarsal	1	1	0	267	4	62
s_metatarsal_d	*	*	*	*	*	*
s_metatarsal_p	*	*	*	*	*	*
s_pelvis	1	0	0	113	2	51
s_phal1	0	*	*	105	0	*
s_phal2	0	*	*	17	*	*
s_phal3	0	*	*	14	*	*
s_phalanges	*	*	*	*	*	26
s_radius	2	3	2	347	*	52
s_radius_d	*	*	*	*	*	*
s_radius_p	*	*	*	*	*	*
s_radius ulna	*	*	*	*	9	*
s_ribs	*	*	*	51	*	*
s_scapula	0	3	0	37	2	21
s_tibia	1	7	3	291	12	95
s_tibia_d	*	*	*	*	*	*
s_tibia_p	*	*	*	*	*	*
s_ulna	0	0	0	93	*	6
s_ulna_p	*	*	*	*	*	*
s_vertebra	0	*	*	164	*	18
SUM NISP	13	38	10	2405	38	503
TOTAL NISP	30	-	-	3010	117	847

SHEEP

Site Name	Meare West 1979 MIA-LIA	Mount Batten total IA	Old Down Farm	Ower LIA-RB	Owslebury	Pennyland MIA
s_astragalus	21	9	28	*	80	*
s_atlas axis	*	6	*	*	*	*
s_calcaneum	25	4	27	*	62	*
s_femur	38	*	90	4	309	10
s_femur_d	*	6	*	*	*	*
s_femur_p	*	5	*	*	*	*
s_humerus	43	*	83	8	304	16
s_humerus_d	*	10	*	*	*	*
s_humerus_p	*	5	*	*	*	*
s_mandible	*	21	158	15	1150	73
s_metacarpal	10	*	81	*	296	*
s_metacarpal_d	*	3	*	*	*	*
s_metacarpal_p	*	6	*	*	*	*
s_metapodials	*	*	*	19	*	18
s_metatarsal	25	*	113	*	445	*
s_metatarsal_d	*	6	*	*	*	*
s_metatarsal_p	*	17	*	*	*	*
s_pelvis	*	10	67	7	173	12
s_phal1	*	12	*	*	183	*
s_phal2	*	2	*	*	85	*
s_phal3	*	3	*	*	28	*
s_phalanges	*	*	180	1	*	4
s_radius	19	*	141	14	492	*
s_radius_d	*	2	*	*	*	*
s_radius_p	*	15	*	*	*	*
s_radius ulna	*	*	*	*	*	31
s_ribs	*	*	85	1	71	7
s_scapula	33	6	43	6	138	14
s_tibia	41	*	130	12	814	32
s_tibia_d	*	5	*	*	*	*
s_tibia_p	*	0	*	*	*	*
s_ulna	*	*	43	3	98	*
s_ulna_p	*	5	*	*	*	*
s_vertebra	*	*	351	7	250	25
SUM NISP	255	137†	1620	97	4978	242
TOTAL NISP	1142	741	-	132	-	341

† For Mount Batten, rather than adding all frequencies, the SUM NISP values were calculated without the lower frequencies for bones for which values for both distal and proximal ends were given

SHEEP

Site Name	Port Seton LIA-ERB	Poundbury total IA	Puckeridge-Braughing LIA-RB	Rope Lake Hole IA-RB	Skeleton Green LIA-RB	Uley Bury
s_astragalus	1	14	5	*	*	0
s_atlas axis	*	*	*	*	*	*
s_calcaneum	1	11	26	*	*	0
s_femur	3	22	124	49	17	22
s_femur_d	*	*	*	*	*	*
s_femur_p	*	*	*	*	*	*
s_humerus	6	27	134	75	39	16
s_humerus_d	*	*	*	*	*	*
s_humerus_p	*	*	*	*	*	*
s_mandible	11	58	208	227	71	96
s_metacarpal	7	22	59	*	21	*
s_metacarpal_d	*	*	*	*	*	*
s_metacarpal_p	*	*	*	*	*	*
s_metapodials	*	*	*	286	*	12
s_metatarsal	8	21	77	*	21	*
s_metatarsal_d	*	*	*	*	*	*
s_metatarsal_p	*	*	*	*	*	*
s_pelvis	4	23	73	44	31	21
s_phal1	0	*	17	*	*	*
s_phal2	1	*	7	*	*	*
s_phal3	0	*	1	*	*	*
s_phalanges	*	61	*	39	7	15
s_radius	5	44	128	163	26	*
s_radius_d	*	*	*	*	*	*
s_radius_p	*	*	*	*	*	*
s_radius ulna	*	*	*	*	*	36
s_ribs	2	100	*	68	*	0
s_scapula	0	37	197	44	38	26
s_tibia	7	35	157	259	41	30
s_tibia_d	*	*	*	*	*	*
s_tibia_p	*	*	*	*	*	*
s_ulna	0	17	33	27	25	*
s_ulna_p	*	*	*	*	*	*
s_vertebra	1	100	3	44	48	0
SUM NISP	57	592	1249	1325	385	274
TOTAL NISP	91	910	1546	-	449	-

SHEEP

Site Name	Wakerley MIA-LIA	Wavendon Gate	West Stow MIA-LIA	Winklebury EIA-MIA
s_astragalus	2	0	6	32
s_atlas axis	*	*	*	*
s_calcaneum	2	0	4	30
s_femur	14	2	19	96
s_femur_d	*	*	*	*
s_femur_p	*	*	*	*
s_humerus	15	5	36	57
s_humerus_d	*	*	*	*
s_humerus_p	*	*	*	*
s_mandible	68	14	85	175
s_metacarpal	18	4	45	69
s_metacarpal_d	*	*	*	*
s_metacarpal_p	*	*	*	*
s_metapodials	*	*	*	*
s_metatarsal	11	4	43	100
s_metatarsal_d	*	*	*	*
s_metatarsal_p	*	*	*	*
s_pelvis	*	4	22	52
s_phal1	3	0	2	84
s_phal2	0	0	1	38
s_phal3	0	0	1	26
s_phalanges	*	*	*	*
s_radius	33	9	66	105
s_radius_d	*	*	*	*
s_radius_p	*	*	*	*
s_radius ulna	*	*	*	*
s_ribs	*	*	*	312
s_scapula	12	1	19	42
s_tibia	40	12	106	171
s_tibia_d	*	*	*	*
s_tibia_p	*	*	*	*
s_ulna	2	3	5	41
s_ulna_p	*	*	*	*
s_vertebra	*	0	7	372
SUM NISP	220	58	467	1802
TOTAL NISP	220	79	890	1802

APPENDIX 2

Frequencies of the 9 skeletal elements of cattle, pig and sheep for the 37 Iron Age sites included in the Principle Component Analysis (PCA). The original data set was taken from Hambleton 1999, appendix 3.

PCA is easily dominated by missing values and, due to the high numbers of missing values in the original faunal data set, its structure had to be altered in order to increase the number of sites that could be used. Three new variables were created: the values for the variables phalanx 1, phalanx 2, phalanx 3 and phalanx are summarised into one variable (x_phal) as phalanx. Metatarsal and metacarpal are combined into a so-called x_foot variable and radius and ulna frequencies are also joined into one variable (x_radul).

Ribs, vertebrae, astragalus and calcaneum were excluded from this analysis due to their high numbers of missing values in the original data set.

SUM $NISP_t$: frequencies of bones for which both species and skeletal element could be identified; frequencies of all three species are summed up in this value
TOTAL $NISP_t$: frequencies of all bones that could be allocated to one of the species (taken from Hambleton 1999, appendix 2); frequencies of all three species are summed up in this value

- missing value

Site Name	Abbotstone Down	Balksbury EIA	Balksbury MIA	Bancroft LBA-EIA	Bancroft LIA-ERB	Bishopstone MIA-LIA
COW						
c_fem	9	9	83	0	7	0
c_hum	20	20	137	4	12	0
c_mand	43	34	217	2	39	13
c_foot	31	22	133	3	31	0
c_pel	12	20	94	0	11	1
c_phal	10	3	19	0	5	0
c_radul	29	25	118	4	23	1
c_scap_p	29	23	123	3	7	1
c_tibia	18	14	105	1	14	0
PIG						
p_fem	4	5	0	0	1	0
p_hum	3	5	28	1	9	0
p_mand	19	10	46	9	13	13
p_foot	1	6	5	2	3	0
p_pel	2	4	8	1	3	1
p_phal	0	1	0	2	2	0
p_radul	5	3	21	2	4	1
p_scap_p	3	4	30	0	10	1
p_tib	6	7	20	1	5	0
SHEEP						
s_fem	3	11	93	0	3	2
s_hum	8	16	96	2	14	9
s_mand	53	39	274	3	27	11
s_foot	42	63	368	5	21	23
s_pel	2	11	80	2	11	2
s_phal	6	8	18	1	3	3
s_radul	21	54	254	6	22	3
s_scap_p	4	8	91	0	9	6
s_tib	40	47	279	0	32	14
SUM $NISP_t$	450	509	3037	58	386	113
TOTAL $NISP_t$	647	764	4132	-	535	629

Site Name	Bramdean	Brighton Hill South	Burgh LIA	Catcote LIA-RB	Cat's Water total IA	Chilbolton Down EIA-MIA
COW						
c_fem	6	37	34	11	137	8
c_hum	2	60	36	34	170	11
c_mand	15	173	107	22	308	8
c_foot	8	76	86	47	325	10
c_pel	4	34	40	6	66	6
c_phal	3	13	23	66	78	5
c_radul	8	63	73	31	226	9
c_scap_p	7	86	55	19	126	9
c_tibia	5	55	52	23	164	3
PIG						
p_fem	1	14	13	1	24	2
p_hum	5	13	14	3	26	1
p_mand	16	50	44	17	57	1
p_foot	4	3	16	8	21	0
p_pel	1	7	24	2	20	0
p_phal	7	8	3	0	1	1
p_radul	4	22	23	0	32	0
p_scap_p	8	27	11	4	39	1
p_tib	2	13	16	1	42	0
SHEEP						
s_fem	8	41	38	13	51	7
s_hum	10	58	38	36	118	5
s_mand	59	181	140	40	253	24
s_foot	10	148	164	40	335	17
s_pel	3	35	28	12	43	6
s_phal	6	29	18	0	25	11
s_radul	8	107	98	42	299	16
s_scap_p	4	30	60	25	67	9
s_tib	14	177	113	33	333	17
SUM $NISP_t$	302	1723	2021	683	3542	200
TOTAL $NISP_t$	-	-	1460	736	5213	357

Site Name	Dalton Parlours MIA-LIA	Danebury EIA	Danebury LIA	Danebury MIA	Danebury MIA-LIA	Ditches LIA-ERB
COW						
c_fem	6	52	65	18	131	92
c_hum	5	78	136	37	200	68
c_mand	13	77	156	39	145	207
c_foot	17	130	166	52	302	110
c_pel	14	81	123	35	169	104
c_phal	42	341	337	140	583	150
c_radul	11	128	261	76	385	99
c_scap_p	11	71	141	46	216	188
c_tibia	7	54	114	40	158	94
PIG						
p_fem	2	36	21	31	79	16
p_hum	2	42	74	42	145	31
p_mand	3	80	149	67	263	90
p_foot	1	100	90	77	239	36
p_pel	2	44	62	40	125	31
p_phal	6	195	161	161	354	22
p_radul	4	104	136	82	278	43
p_scap_p	0	55	81	37	174	34
p_tib	1	47	66	35	112	22
SHEEP						
s_fem	18	143	259	95	433	66
s_hum	8	263	588	142	782	90
s_mand	12	383	1033	241	1510	206
s_foot	54	380	700	219	1060	86
s_pel	9	200	486	124	614	126
s_phal	239	728	967	480	1194	23
s_radul	31	385	758	199	1038	131
s_scap_p	8	182	441	106	677	161
s_tib	23	246	489	117	626	173
SUM $NISP_t$	549	5378	9459	3248	11869	2706
TOTAL $NISP_t$	695	15207	26405	9322	39585	4340

Site Name	Dragonby MIA-LIA	Farningham Hill LIA	Grimthorpe IA	Maiden Castle EIA-LIA	Meare 1984 LIA	Mount Batten total IA
COW						
c_fem	86	13	11	41	18	5
c_hum	78	19	17	45	17	4
c_mand	150	41	31	71	49	12
c_foot	159	28	53	100	20	39
c_pel	103	23	22	35	5	13
c_phal	164	15	29	85	7	82
c_radul	186	21	30	96	35	13
c_scap_p	124	30	13	43	8	17
c_tibia	104	16	19	41	23	9
PIG						
p_fem	37	0	1	8	18	0
p_hum	65	6	8	30	14	9
p_mand	120	7	7	36	19	72
p_foot	33	0	1	17	7	24
p_pel	21	6	3	18	15	8
p_phal	28	3	1	21	24	33
p_radul	63	2	12	42	4	12
p_scap_p	73	6	3	42	9	22
p_tib	52	3	1	15	17	5
SHEEP						
s_fem	185	7	10	171	29	6
s_hum	243	11	5	210	37	10
s_mand	455	28	26	286	52	21
s_foot	425	11	12	446	101	23
s_pel	161	8	17	113	51	10
s_phal	79	14	2	136	26	17
s_radul	366	27	12	440	58	20
s_scap_p	209	5	0	37	21	6
s_tib	488	31	13	291	95	5
SUM $NISP_t$	4407	457	438	3471	840	639
TOTAL $NISP_t$	8718	1095	644	4365	1487	2945

Site Name	Old Down Farm	Ower LIA-RB	Owslebury	Pennyland MIA	Port Seton LIA-ERB	Poundbury total IA
COW						
c_fem	36	3	244	38	14	8
c_hum	68	4	306	43	23	15
c_mand	75	12	1084	68	89	18
c_foot	62	9	510	64	64	33
c_pel	73	8	253	40	36	7
c_phal	64	5	236	37	15	11
c_radul	79	7	409	52	35	21
c_scap_p	38	0	444	31	30	23
c_tibia	48	4	285	57	22	15
PIG						
p_fem	4	9	143	1	0	8
p_hum	16	4	206	8	4	4
p_mand	40	41	685	16	25	20
p_foot	15	3	69	3	3	4
p_pel	4	4	97	3	1	5
p_phal	12	4	150	1	0	1
p_radul	22	6	157	6	4	6
p_scap_p	9	7	167	8	4	10
p_tib	10	8	174	8	0	4
SHEEP						
s_fem	90	4	309	10	3	22
s_hum	83	8	304	16	6	27
s_mand	158	15	1150	73	11	58
s_foot	194	19	741	18	15	43
s_pel	67	7	173	12	4	23
s_phal	180	1	296	4	1	61
s_radul	184	17	590	31	5	61
s_scap_p	43	6	138	14	0	37
s_tib	130	12	814	32	7	35
SUM $NISP_t$	2490	252	11436	845	477	941
TOTAL $NISP_t$	-	471	-	1145	707	1462

Site Name	Puckeridge-Braughing LIA-RB	Rope Lake Hole IA-RB	Skeleton Green LIA-RB	Uley Bury	Wavendon Gate	West Stow MIA-LIA
COW						
c_fem	63	21	46	8	40	51
c_hum	117	40	62	19	25	54
c_mand	88	103	42	64	74	127
c_foot	150	98	51	18	16	123
c_pel	93	40	41	10	23	38
c_phal	131	62	49	29	4	93
c_radul	142	51	73	19	42	83
c_scap_p	139	46	77	19	32	69
c_tibia	132	25	64	21	36	76
PIG						
p_fem	76	2	43	0	0	9
p_hum	85	10	49	5	0	28
p_mand	171	30	229	47	3	34
p_foot	93	12	92	3	0	12
p_pel	85	4	75	6	4	15
p_phal	26	6	24	2	0	5
p_radul	149	16	94	6	1	10
p_scap_p	103	16	65	23	1	26
p_tib	140	10	73	7	0	3
SHEEP						
s_fem	124	49	17	22	2	19
s_hum	134	75	39	16	5	36
s_mand	208	227	71	96	14	85
s_foot	136	286	42	12	8	88
s_pel	73	44	31	21	4	22
s_phal	25	39	7	15	0	4
s_radul	161	190	51	36	12	71
s_scap_p	197	44	38	26	1	19
s_tib	157	259	41	30	12	106
SUM $NISP_t$	3382	2043	1875	670	379	1421
TOTAL $NISP_t$	4306	-	2437	-	503	2550

Site Name	Winklebury EIA-MIA
COW	
c_fem	28
c_hum	20
c_mand	69
c_foot	44
c_pel	59
c_phal	45
c_radul	39
c_scap_p	47
c_tibia	36
PIG	
p_fem	8
p_hum	18
p_mand	40
p_foot	0
p_pel	8
p_phal	18
p_radul	50
p_scap_p	36
p_tib	26
SHEEP	
s_fem	96
s_hum	57
s_mand	175
s_foot	169
s_pel	52
s_phal	148
s_radul	146
s_scap_p	42
s_tib	171
SUM $NISP_t$	2813
TOTAL $NISP_t$	2817

www.ingramcontent.com/pod-product-compliance
Lightning Source LLC
LaVergne TN
LVHW071216210926
848482LV00015B/823

* 9 7 8 1 4 0 7 3 0 5 8 7 5 *